PRODUCE

A Fruit and Vegetable Lovers' Guide

PRODUCE

A Fruit and Vegetable Lovers' Guide
Written by Bruce Beck
Photographed by Andrew Unangst
Illustrated by Rodica Prato

The publishers wish to thank Balducci's
and their produce manager, Dominick
Doria, for both technical assistance
and the selection of fruits and vegetables
photographed in this book.

Published by
Friendly Press, Inc. 401 Park Avenue South,
New York City 10016, United States of America

Printed in Italy by Arnoldo Mondadori Editore Verona

Design—John Carrod/Black Book Studios

Photo Coordinator—Nancy Weaver

Library of Congress Cataloging in Publication Data

Beck, Bruce.
Produce: a fruit and vegetable lovers' guide.

Includes index.
1. Fruit.   2. Vegetables.   3. Marketing (Home economics)
I. Title.
TX397.B44   1984   641.3′4   83-49410

ISBN 0-914919-01-6

ACKNOWLEDGMENTS

My deepest gratitude must go to Stu Waldman, who conceived this book, and whose care and guidance have made it what it is, and to Peggy Flaum and Marty Goldstein, who complete the triumvirate that worked and worried over this project from the beginning. Special thanks go to the entire staff of Friendly Press for their warm interest, particularly to Nancy Weaver for her untiring efforts to keep this book on schedule and outwit the weather to ensure photo props; to Sharon Kaplan and Meggin Chinkel Siefert for their meticulous attention to getting the book printed beautifully; to John Carrod for his tasty designing; and to Maggie Groening for her invaluable editorial assistance.

An astonishing majority of the produce for the photographs was provided by Dominick Doria, whose enthusiasm and experience were just as important to this book as they are to Balducci's Gourmet, one of the finest markets in the U.S. Mr. Doria's knowledge of his craft was an inestimable aid.

The rigorous photography schedule could not have been met without the able cooperation of the Unangst Studio staff: Robin Gneiting, manager; Doug Whyte, assistant; Deborah Hansen, stylist; and Kimberli Lynch, secretary.

I wish to thank my friend and colleague Norman Weinstein for his help with Chinese vegetables, and my friend June Sabah for sharing her knowledge of produce Caribbean and Middle Eastern. Important thanks go to John Lowy, Director, and William Liederman, Executive Director of The New York Cooking Center, for their faith in me, and for giving me the scheduling flexibility needed to complete this book. A final thank you goes to all my students, for what they have taught me, and for what they have made me learn.

to William Roy
without whose patience and support the
writing would not have been possible

CONTENTS

INTRODUCTION

Recently I happened to observe a woman of a certain age, a woman with the air of an experienced shopper, peruse the selection of pineapples in the supermarket, select one for testing, pluck a leaf from the crown, smile a tiny, private smile of approval, and confidently place the fruit in her shopping cart, secure in the knowledge (so I imagined) that she had chosen wisely for her family and her budget. She proceeded to honeydew melons, and tested a few by shaking them vigorously close to her ear, in the hope of detecting just the right sound. Number three passed the test and it too went carefully into her cart. This shopper was undeterred by the fact that the pineapple was quite small and green as grass, or that it was April, when fine sweet honeydews are rare as hens' teeth. She was also blissfully ignorant of the invalidity of her tests. Politeness and other obligations prevented me from studying her further, but she left a strong impression of exactly why this book was undertaken.

Most shoppers seem not to know how to select fine quality produce, and many grocers are hardly better informed. This situation is unfortunate under any circumstances, but especially now, when there is more interest and variety than ever before. Health consciousness and 'the new cooking' have spurred an exciting trend, with modern storage and transportation techniques making possible an abundance and selection of fresh produce never before known. The other twentieth century trend persists—standardization to the dullest common denominator and weeding out the weak—but the public has made it clear that it wants variety, and the industry has responded, as it always does. Mammoth growers and tiny greenhouses alike answer the call, as do enterprising retailers.

I sincerely hope that this book will have a positive effect on the produce industry by arming consumers with the information they need to make intelligent purchases. Sales figures are the voice of the people, and I wish that voice to make a cry for quality and freshness, as well as variety. Experience has shown that instead of getting what we pay for, we get what we will buy.

COMMERCIAL APPLES

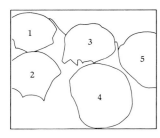

1. McIntosh – the third largest crop. This is a pleasant, all-purpose apple that can be surprisingly good when bought in the fall from local grocers. Available most of the year, least in mid-summer.

2. Granny Smith – a new variety from New Zealand now grown in the U.S. and France. This is the only superior apple on the mass market for eating and cooking. Available fresh year round.

3. Red Delicious – the most popular apple, also among the most insipid. Available year round.

4. Golden Delicious – golden, yes, delicious, no. Second in popularity with even less flavor than Red. Year round.

5. Rome Beauty – the smallest crop (and the largest fruit) of the supermarket apples. Mainly a cooking apple. Available all but August and September.

More than just the largest fruit crop in temperate climates, the apple is, to the Western mind at least, the archetype for all fruits. Throughout history dozens of things with a vaguely spherical shape have been dubbed apples until a more individual name was found. Some never really got new names, so that in modern times we still have *pomme de terre* ('earth apple,' French for potato), pomegranate (also called Chinese apple), *pomidoro* (Italian for tomato, from *pomo d'oro*, 'golden apple,' also called love apple, *pomme d'amour* in French), custard apple, pineapple, and others. And it need not be a fruit for the name to stick – witness road apple (a vanishing phenomenon because of the automobile), Adam's apple, and the ultimate term of respect, the Big Apple. The sole fact that the biblical fruit of the tree of knowledge of good and evil became an apple in art and popular culture is enough to establish this member of the rose family as *the* Western fruit.

Harvest time is late summer to late fall, but many varieties store well into winter. There is a modern treatment involving storage in a special oxygen-free environment that extends the shelf life of some varieties for months. Flavor and texture do not suffer badly, because these are varieties with little to begin with. Choose fruit that is bright skinned, plump, and crisp-looking. Avoid bruises and rot, for they spread rapidly. Shriveled fruit is of no interest for eating, but it is useful for hastening the ripening of other fruits, using the paper bag method: simply add a very ripe apple to the bag with the other fruit, be sure to punch a few holes in the bag, and the natural ethylene gas released by the ripe apple will speed things along.

Apples are perhaps the most versatile of fruits. They can be baked, stewed, poached, broiled, puréed (applesauce), fried, sautéed, pressed for cider and brandy, and, of course, eaten out of hand. Apples work well in salads, pies, tarts, fritters, cakes, relishes, jellies, preserves, and mincemeat. The flavor and texture complement, or are enhanced by, raisins and dried currants, citrus fruits, nearly any other fruit, sweet spice, brandy, liqueurs, wine, brown sugar and molasses, pork, veal, chicken, duck and goose, turkey, game, and dairy products, especially cream, ice cream, and cheese. Apples are traditional in a variety of savory concoctions, including sauerkraut and holiday bread stuffings.

WHAT TO LOOK FOR: bright-skinned, plump, crisp-looking apples

WHAT TO AVOID: bruises, rot, shriveled skin; lightweight fruit

HOW TO STORE: refrigerate for a week or two, or store in a cool, dry, dark, well-ventilated place for longer periods

PRIME SEASON: (Indicated by darker shade).

| JAN | FEB | MAR | APR | MAY | JUN | JUL | AUG | SEP | OCT | NOV | DEC |

LOCAL APPLES

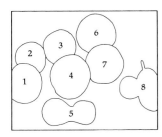

1. Paula Red – an obscure, early-ripening New England apple with a charming, clear red blush. Pleasant but undistinguished.

2. Gravenstein – an early-ripening sauce or eating apple that rarely sees the market, except in New England and California.

3. Macoun – a McIntosh hybrid of very high quality. The fall season is short and Macouns do not keep well. One of the best.

4. Empire – a McIntosh hybrid, handsome and red, crisp, pleasant.

5. Lady – a charming little red and yellow thing, perfect for eating out of hand. This is the Api, a very old variety much prized in Europe and colonial America.

6. Cortland – yet another McIntosh hybrid, pleasantly crisp, not very tangy, good for salads because it browns slowly once cut.

7. Greening – *the* traditional pie apple.

8. Crab – the beautiful little round red apple, too tart to eat raw, prized for pickles and preserves.

Names such as Jonathan, Winesap, Stayman, Baldwin, York Imperial, Grimes Golden, Wealthy, Northern Spy, and Newtown Pippin conjure up a vision of Johnny Appleseed and a nostalgic American past replete with crisp autumn days, crisper sweet/tart juicy apples, fresh-pressed cider, and, of course, Mom's apple pie. Unfortunately, nostalgia it is, for most of these varieties are simply not available to the average consumer. Some of them may not be grown commercially at all by the time this book goes to press. Others are processed immediately and never make it to market. Still others do not meet modern packaging, shipping, and shelf life demands, so they are not transported any distance or are being phased out.

Luckily, the situation in Europe is not so grim, though standardized U.S. apples are making strong inroads. Red and Golden Delicious are now impor-

tant European crops, but it is still possible to find Worcester Pearmain, Cox's Golden Pippin, Calville, Duchess, Cellini, Annurca, Rennet (Reinette), some of the varieties in the previous paragraph, and a number of other fine apples in good supply.

The only reliable way to find honest apples is to live near an orchard, or journey to one, and to know just when a favorite variety is coming to harvest. The other possibility is to seek out a good market that stocks some of the rarer varieties, then enjoy them during the fall season. Given a cellar, one can store the hardiest varieties into the winter. Fine quality apples have the same uses as the mass-market ones; they just taste better.

WHAT TO LOOK FOR:
bright-skinned, plump, and crisp-looking apples

WHAT TO AVOID:
bruises, rot, shriveled skin; lightweight fruit

HOW TO STORE:
refrigerate for a week or two, or store in a cool, dry, dark, well-ventilated place for longer periods

PRIME SEASON: (Indicated by darker shade). This may vary slightly due to local crops.

JAN   FEB   MAR   APR   MAY   JUN   JUL   AUG   SEP   OCT   NOV   DEC

APRICOTS

If only they tasted as lovely as they look. Well, sometimes they do. Your best bet is to live next to an apricot orchard in California, Turkey, Spain, Greece, or maybe New Zealand. Then you wait for the fruit to get soft and luscious in midsummer and you make your move. It is the old transportation problem – fully ripened apricots are so fragile, they will not even travel well across the street. And only in California will you find the absolute best apricots for eating fresh. The others are finer dried or put up in syrup or preserves. The closer you live to the

growing area, the more tree-ripened the available apricots will be, but even thousands of miles away, careful shoppers can still find a taste of the summer sun.

The best fresh-market apricot in most areas is called Royal. Royals are medium to large, rich yellow with a bit of blush, and decidedly round with only a minor indentation on one side. They are freestone. Avoid the others. Choose fruit of proper color that is apricot scented and beginning to soften. Reject any that are pale or rock-hard and, of course, any that are bruised, shriveled, or rotting. Handle with care and complete the ripening at room temperature in a paper bag with a few holes in it. Use promptly once they are very soft, golden, and sweet-smelling. Outside the growing areas, apricots are usually available only in June, July, and maybe August. Air shipments from New Zealand show up in January, February, and maybe March.

Because it takes care to choose and mature perfect apricots, it seems only fair to enjoy them naked and unadorned. But apricots do blend nicely with other fruits. They also enhance poultry in salads and hot dishes. Nuts complement them well, as does lemon juice, and apricots can be used for tarts, preserves, and relishes, as well as ices and ice creams.

WHAT TO LOOK FOR:
Royals – medium to large, round, yellow with blush, freestone, apricot scent

WHAT TO AVOID:
pale or greenish, rock-hard fruit, bruises, shrivel, rot

HOW TO STORE:
ripen at room temperature in a paper bag with a few holes in it; use promptly

PRIME SEASON: (Indicated by darker shade).

| JAN | FEB | MAR | APR | MAY | JUN | JUL | AUG | SEP | OCT | NOV | DEC |

ARTICHOKES

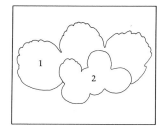

1. Large
2. Tiny

These artichokes are in typical, acceptable condition. In markets close to growing areas it is possible to find them in an even finer, less blemished state (provided they have not been 'winter-kissed').

'Remind me to tell you about the time I looked into the heart of an artichoke.' So said Bette Davis's Margo Channing in the film *All About Eve.* This was pretty sophisticated stuff for 1950, when most Americans had never seen an artichoke, let alone peered into the heart of one. Luckily, that has changed. The cool, foggy coastal area around Castroville, California, now turns out an annual crop to rival the traditional yield from the south of France, Spain, and Italy. It is often called globe artichoke, and if not picked young, it will grow up to be a big, purplish, thistlelike flower. But artichokes are nipped in the bud to produce one of the most interesting vegetables around. They have a curious flavor, subtle, nutty, and green. And the most curious part is that they alter the chemistry of the taste buds to make other foods taste sweeter, a property that bedevils wine purists and charms everyone else.

Artichokes are available year round, with a peak, and the lowest prices, in spring. Whether they are very round or more cone shaped, they should always be tightly headed. Spreading leaves are a sure sign of staleness and flavor loss. Choose artichokes that feel heavy and that have fresh green leaves and stems. Size is not a quality factor. In addition to spread, avoid softness and rust or blackening. The one exception is cold weather artichokes

that have been slightly chill-damaged. 'Winter-kissed' artichokes will be darkened a bit at the leaf tips and are perfectly fine so long as they are in good condition otherwise. Use promptly or refrigerate for a day or two.

Though sometimes stuffed, baked, or fried, artichokes are often steamed or simmered until tender and served with a dipping sauce–lemon butter or *sauce hollandaise* for hot and vinaigrette or mayonnaise for cold. Diners pull off the leaves one at a time, dip them, and scrape off the flesh between the teeth. Eventually one comes to the pale whitish and purplish center leaves, which are spooned away and discarded along with the fuzzy 'choke.' Many Italian artichokes have edible centers, but the others must be removed to avoid unpleasant stinging and irritation. Now the fat, fleshy bottom is revealed, the prized part that is eaten with knife and fork. Artichokes are sometimes trimmed down to the bottom before cooking, yielding a dish shape that is usually filled with something. They can also be trimmed of outer leaves and neatened up to yield an artichoke heart. To prevent discoloration, they should be cut with a stainless steel knife and immediately acidulated with lemon juice or vinegar. Artichokes team well with herbs, vegetables, eggs, cheese, butter or olive oil, lamb, chicken, and seafood, especially crab and shrimp. They are enhanced by highly seasoned or delicate sauces.

WHAT TO LOOK FOR:
firm, heavy, green globes

WHAT TO AVOID:
spreading leaves, rust or blackening (except for 'winter-kiss'), rot or flabbiness

HOW TO STORE:
refrigerate for a few days

PRIME SEASON: (Indicated by darker shade).

| JAN | FEB | MAR | APR | MAY | JUN | JUL | AUG | SEP | OCT | NOV | DEC |

ARUGULA

The good word is that rocket is back after a long absence and it has a new name – arugula. By the turn of the century it had all but disappeared from gardens and greengrocers everywhere save Italy. But arugula has come into its own, along with so many other glories of the Italian kitchen. Rocket is the proper English word, most likely from the shape of the leaf tip. It has the same name in France – *roquette*. In Italy it is called *rucola* or *rugala*, but arugula (or arugala) seems to be a wholly American version of the Italian name. By any name, it is pungent and peppery with a distinctive quality that reminds some of horseradish and others of mustard. (It is a member of the mustard family.)

Arugula is available most of the year from greengrocers who are well stocked. Expect to find it absent from time to time in winter, and at its best and most plentiful in summer. For the most delicate flavor and texture, choose very fresh small leaves with good green color. As leaves grow older and larger, the characteristic pungency increases markedly. Avoid yellowed, blighted, or wilted leaves. If necessary, arugula may be stored for a day or two, refrigerated. Home gardeners find it easily grown from seed.

A salad of pure arugula is expensive and a bit too highly flavored for most tastes. Arugula is especially good in combination with blander greens such as Boston and Bibb lettuce. The flavor marries well with tomatoes and scallions.

Wash gently but thoroughly by immersing in several changes of cold water until there is no trace of grit. Dry carefully with clean towels – the leaves are tender and easily bruised. Refrigerate, wrapped in a towel, for an hour or two until needed. Use a dressing made with the finest quality olive oil and wine vinegar. Extra-virgin olive oil from Italy is ideal.

Arugula is especially refreshing when it accompanies or follows lamb, rich or highly seasoned foods, and the fatty flesh of duck or pork. In summer – or anytime – it is perfect with grilled or barbecued meats or seafood.

WHAT TO LOOK FOR:
fresh, good green color

WHAT TO AVOID:
overgrown leaves, yellowing, blight, wilt

HOW TO STORE:
refrigerate wrapped, unwashed, for a day or two

PRIME SEASON: (Indicated by darker shade).

| JAN | FEB | MAR | APR | MAY | JUN | JUL | AUG | SEP | OCT | NOV | DEC |

ASPARAGUS

1. White
2. Green

One of the most prized and elegant of all the world's vegetables is this young shoot of a plant that grows up to be a big, feathery, fernlike thing. It is a cousin of the onion and other lilies, though it bears no resemblance to them in appearance or flavor. Sometimes called sparrowgrass, or grass (in the produce industry), or even housemaid's horror, asparagus has enjoyed a zealous following since ancient times, first from Britain to the Mediterranean, then throughout the world. Wild asparagus differs very little from the cultivated version, although purists find it more desirable. Asparagus has been eaten for its alleged aphrodisiacal and medicinal properties, but mostly it is eaten for its suave flavor, a fresh greenness that is delectable.

Asparagus is one of the major glories of springtime, at its best when harvested from cool foggy fields and rushed from market to table. It is only perfect when very fresh, for its natural sugars change with age, converting to woodiness and throwing off taste and texture. For this reason, the crops available at other times of the year from other areas are likely to be less than stellar, especially when you consider their great expense. Asparagus comes in both green and a milder (and more expensive) white form, the white having been blanched by being covered with soil. Choosing between the two is strictly a matter of preference when freshness is equal. Whether the stems are fat or skinny, they must be plump and crisp with tightly furled tips. Spreading tips and wilted or shriveled stems are indications that the asparagus is past its prime. If necessary, it may be wrapped and refrigerated for a day or two.

Too few people realize that asparagus is a delightful raw vegetable, tasty and attractive when served as *crudité* with an interesting dipping sauce. Whether steamed or simmered, peeled or unpeeled, cut uniformly or snapped at the point of tenderness, asparagus should always be cooked lightly, just until the stem wilts a bit. It goes well with any meat and is at its simple best when dressed only with melted butter and egg sauces, or black butter and capers. And there are possibilities for *gratins*, herbal additions, and even Asian treatments. As salad, asparagus should always be sauced at serving time, for the acid in most dressings will turn it yellow.

WHAT TO LOOK FOR:
plump, crisp, straight stems, tightly furled tips

WHAT TO AVOID:
spreading tips, wilted or shriveled stems

HOW TO STORE:
refrigerate wrapped, for a day or two

PRIME SEASON: (Indicated by darker shade).

JAN	FEB	MAR	APR	MAY	JUN	JUL	AUG	SEP	OCT	NOV	DEC

AVOCADOS

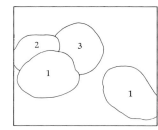

1. Hass (Mexican)
2. Fuerte (Guatemalan)
3. Florida (West Indian)

Would you take an alligator pear to lunch? How about the fruit of the testicle tree? They are one and the same, and they come to lunch and other meals under the name avocado. The fruit is often pear shaped, but the alligator part is a bit vague, coming either from the rough skin of some varieties or from one of many attempts to transliterate, interjecting something familiar, the Aztec name *ahuactl.* This Aztec name is the earliest one and it really means 'fruit of the testicle tree,' probably because of reputed aphrodisiacal qualities, shape, and the fruit's habit of growing in pairs. And there are other names as well, including poor man's butter, midshipman's or subaltern's butter, and even love fruit. It is a very popular fruit these days, with soft, buttery flesh that is slightly nutty, very soothing, and remarkably rich in oil (and calories–about 275 for half a medium-sized avocado).

Avocados are available year round with the largest crop of the best type–Hass, or Mexican–in the market from January through May. Choose firm, attractive fruit that is fresh-smelling, free from bruises or cuts, and that yields, at least a bit, to

gentle pressure. Rock-hard avocados have been picked too green and will probably rot before they ripen. The chances of finding a perfectly ripe avocado in the market on the day you want to use it are poor, so plan ahead a few days and ripen it at home using the bag technique. A perfectly ripe avocado will yield easily to the slightest pressure without being mushy in the least. Once cut, avocados brown rapidly, so brush the surfaces with lime or lemon juice and wrap airtight. An unused half should be treated as above, refrigerated, with the pit in place. The pit is reputed to have anti-browning effects.

The most famous presentation of avocados is the Mexican-American salad or dip called guacamole. For salad purposes, avocados marry well with greens, sprouts, chiles, fresh herbs, onions, garlic, lemon or lime, tomatoes, breast of chicken or turkey, nuts, eggs, bacon, shellfish, and vinaigrette or mayonnaise. Avocado halves are sometimes stuffed or peeled and cut into fans. Purée of avocado makes a satisfying spread for breakfast toast and, sliced or diced, the fruit has many sandwich possibilities. It is all the rage in sushi parlors from New York to Tokyo teamed with crab and rice and wrapped in seaweed, usually called a California roll. Very slight cooking is essential when avocados are baked or steamed to be served as a hot vegetable. They also turn up in soups and even ice cream.

WHAT TO LOOK FOR: firm, fresh-smelling fruit

WHAT TO AVOID: bruises, cuts, rock-hard or mushy fruit

HOW TO STORE: ripen at room temperature, then refrigerate for a few days if necessary

PRIME SEASON: (Indicated by darker shade).

| JAN | FEB | MAR | APR | MAY | JUN | JUL | AUG | SEP | OCT | NOV | DEC |

BANANAS AND PLANTAINS

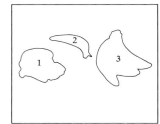

1. Red bananas
2. Plantain
3. Yellow bananas

Though the red bananas are a bit more bruised than is ideal, each type is in proper condition for purchase. The yellow bananas are about as ripe as they should ever be in the market (to avoid bruising).

If there is perfection in the modern world, then the banana may be it. No tropical fruit is so ideally suited to the demands of modern agriculture, for–and here is the zinger–it must ripen off the plant. Firm green fruit travels safely in the holds of air-conditioned ships, especially after the cargo is gassed to eliminate tarantulas and pythons not wanted on the voyage. Ripening is started in the warehouse; when it gets to market, the grocer simply puts the bunches out and the rest of us wait for them to come around to melting sweetness. No mess, no fuss, no poking, squeezing, sniffing, shaking, thumping. Provided we do not chill it (never put unripe bananas in the refrigerator), it will move predictably along from greenish to yellow with brown speckles called honey spots, a reliable sign of ripeness. Through nature and mankind working together, bananas can be enjoyed anywhere in the world at their peak of tropical goodness. The same is true for the banana's large starchy cousin plantain, a fruit that shares the distinctive flavor and aroma of bananas but never develops sweetness for eating as a fruit. Plantains require heat to bring out their subtle vegetable soothingness.

Though plantains are less popular and harder to find than bananas, both are in steady supply all year. Bananas should be purchased while still firm, unblemished, and showing a bit of green. Avoid black bruises and mushiness. Ripen at home at room temperature to the desired softness, then refrigerate if necessary. Plantains should be purchased when deep green, when they are firm and starchy. They may be ripened (becoming yellowish, then decidedly black) at room temperature if desired. When peeling bananas or plantains, be sure to remove the bitter strings that run the length of the flesh at about half-inch intervals. In addition, green plantains should be peeled under cold running water to avoid black stains on the hands.

Bananas are delicious eaten out of hand, in fruit salads, pies, cakes, puddings, breads, fritters, soufflés, custards, curries, milk shakes, and, of course, banana splits. They can be used raw, baked, broiled, or fried, and they team well with chocolate, coconut, any fruits, eggs and dairy products for puréed breakfast drinks, and even pork and poultry. There are also possibilities for chutneys and preserves. In addition to the familiar large yellow banana, there are tiny yellow and sometimes even red ones on the world market that are very tasty and sweet. Plantains are always cooked, whether green or ripe. They are most often boiled, broiled, baked, or fried, to be served as a starchy vegetable with roast meats and beans and rice. Plantains make delicious chips for snacking, and are sometimes sweetened. More often they are served plain, perhaps with lemon or lime juice, or mixed with vegetables–onions, garlic, tomatoes, squashes, peppers, and the like.

WHAT TO LOOK FOR:
firm, unblemished, slightly green bananas; deep-green plantains

WHAT TO AVOID:
dark bruises, mushiness

HOW TO STORE:
ripen at room temperature, then refrigerate if necessary

PRIME SEASON: (Indicated by darker shade).

| JAN | FEB | MAR | APR | MAY | JUN | JUL | AUG | SEP | OCT | NOV | DEC |

BASIL

The little flower buds at the end of the stem are common but not ideal (always pinch them back if you grow your own). Before buying basil which has begun to flower, taste a leaf to be sure the flavor is clean and strong but not bitter.

If it is true that an army marches on its stomach, then the vast army of New Yorkers seems to be marching on basil these days. This kingly herb (from the Greek *basilikón*) has always been in the spice rack, dried, to be sprinkled sparingly, but recently Americans have started following the example of Italians, lavishing fresh basil about in generous quantities.

Since ancient times, peoples of the Mediterranean and India have prized this pungent member of the mint family as food, medicine, breath freshener, room deodorizer, love token, aphrodisiac, insect repellent, safe passage for souls of the dead, and religious symbol, pagan and Christian. Not all has been praise, however, for some medieval physicians warned that the mere scent would breed scorpions in the brain. It is a heady scent, heralding a warm, rich, minty flavor.

Fresh basil is now grown year round in greenhouses, augmenting the usual summer crop. Authorities believe that the small-leafed Italian basil has the finest flavor, but elsewhere the available crop is likely to be large leafed and still delicious.

There is also an attractive variety with a deep purple color, sometimes available in specialty markets. Look for fresh, brightly colored leaves. Avoid rust, wilt, yellowing, and flower buds. Flowering is an indication that the basil is overgrown or ill-tended and will probably be unpleasantly bitter. Wrap in damp paper and refrigerate for a day or two if necessary, but do not wash until ready to use (and then very gently), for wetness encourages decay.

Basil is most closely linked to tomatoes, raw and in sauces, and they do blend beautifully. Another famous use is the Genovese paste called *pesto*–large quantities of basil pounded with garlic, Parmesan and Romano cheeses, olive oil, butter, and pine nuts. This is a delicious sauce for pasta and a flavoring agent for soups, as is the similar *pistou* of southern France. Basil perks up other soups as well as stews, casseroles, curries, and salads. It often joins garlic and/or olive oil to add savor to veal, poultry, lamb, pork, fish, and shellfish. In addition to tomatoes and garlic, basil teams well with onions, eggplant, peppers, zucchini, other herbs– in short, almost anything Mediterranean.

WHAT TO LOOK FOR:
freshness, bright color

WHAT TO AVOID:
wilt, rust, yellowing, flower buds

HOW TO STORE:
refrigerate unwashed, wrapped in damp paper, for a day or two

PRIME SEASON: (Indicated by darker shade).

| JAN | FEB | MAR | APR | MAY | JUN | JUL | AUG | SEP | OCT | NOV | DEC |

Beans

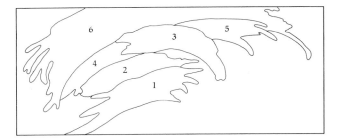

1. Green bean
2. Cranberry bean
3. Fava
4. Tiny green bean (Haricot)
5. Wax bean
6. Chinese long beans

Legume is the prosaic name for some of the world's most important foods. It covers all the podded plants, and beans are among the most varied of these and any other plants. For sheer diversity, it is tough to beat red, yellow, green, purple, beige, brown, tan, pink, black, white, and dappled. And that's just color. The range of shapes and sizes is impressive as well. There are beans that are consumed in their entirety and those with tough pods that are discarded once the seeds are removed. Many beans are grown exclusively, or nearly so, for drying. But the fresh-market supply is still interesting, plentiful, and delicious.

NEW WORLD BEANS, OR HARICOTS

These are the ones that the earliest European explorers discovered growing in a wide range of climates in the Americas. They gained great popularity back in Europe, especially France and Italy, and have been extensively bred and cultivated in both worlds. For fresh market, the most popular ones are green beans of various shapes and sizes–tiny French ones; medium-sized ones; fat ones of American, English, and Italian development, usually called pole beans; and green beans that happen to be yellow, usually called wax beans. All of these will mature to produce fat seeds and tough, inedible pods, but immature pods are normally used. The notable exception is Scarlet Runner, the most popular bean in Britain, which is used both ways. The medium-sized ones, called string beans (and sometimes stringless beans, a more accurate name for most modern varieties) or snap beans, are available all year, at their best and most plentiful in summer. The others are also most available in summer, and sporadically the rest of the year. Choose plump, velvety beans of even color. The smaller ones in any group are usually more tender. The pointed blossom end should be fresh. In addition to shriveled ends, avoid shrivel anywhere else and wilt, rust, overgrown specimens, or swelling around large seeds, except in pole beans, where this is normal (within reason) and the large seeds are tender. Green beans can be bagged and refrigerated for a day or two if necessary.

Only the stem end should be removed, and green beans retain their food value best when they are eaten raw or cooked whole, and briefly. They go with any main-dish meats and are successful in soups, stews, salads, casseroles, *gratins*, and vegetable combinations. They are sometimes fried, stir-fried (after blanching), or pickled.

Soybeans are the most important beans in the world, though they are rarely used fresh. The U.S. produces most of the world crop. The Chinese diet is based upon them, and so is the Western diet, in a way, if we consider the oil, meal, and livestock feed they provide. If you happen to find them fresh, you might want to cook them but you probably won't like them. Chickpeas are similarly important in the Middle East and India, and the rare fresh ones can be used like the dried. In growing areas, especially the American South, the various beans called black-eyed peas, cow peas, and crowder peas are occasionally to be found fresh, but dried ones are more common. Fresh pigeon peas can sometimes be found in Latin markets, but do not confuse them with the mushy frozen beans that have been defrosted.

PRIME SEASON: (Indicated by darker shade).

| JAN | FEB | MAR | APR | MAY | JUN | JUL | AUG | SEP | OCT | NOV | DEC |

The other two most popular fresh New World beans are cranberry beans, of which only the seeds are eaten, and lima or butter beans, whose very young pods can be eaten but rarely are. These two are less available fresh than green beans and are best in summer. Choose and store them as you would other beans. Limas come in a small variety called baby limas, and the large one called Fordhook. Though both are mild and tender, Fordhooks are less so, and each has its devotées. Cranberries and limas should be shelled just before cooking, and they can be treated as fresh vegetables or cooked and served in preparations where dried beans are traditional. Beware of 'fresh' shelled beans in bags, which are actually dried beans that have been soaked and dyed. Dried or frozen beans are better choices. Limas must always be thoroughly cooked to minimize their natural cyanide.

BROAD BEANS

Broad beans, or favas, are important fresh and dried, very popular from England to Africa to the Middle East. They grow in very fat green pods that look as if they would be hard but are actually quite soft. Very young pods can be eaten, but more often only the seeds are. Older pods can produce a condition called favism, which is variously described as a hereditary allergic reaction or the result of excessive consumption. It seems that anyone can be poisoned by undercooked broad beans. Fresh broad beans are available sporadically, especially in spring and summer. They are a starchy staple used in many ways as hot and cold appetizers, in soups and vegetable dishes, and as accompaniments to meats, especially lamb and cured pork.

CHINESE LONG BEANS

Dau gok are available on a regular basis in Asian markets. They look like very long string beans and can grow to be a yard long, though more often just over a foot. They should be chosen and stored in the same ways as their Western cousins. Long beans have a nice crispness but they are considerably more tender than string beans and can be stir-fried without blanching. They might also be substituted in Western recipes.

WHAT TO LOOK FOR:
plump, velvety beans, even color, small for their type, fresh blossom ends

WHAT TO AVOID:
shrivel, rust, wilt, overgrowth, especially overdeveloped seeds in types that are supposed to be immature

HOW TO STORE:
bag and refrigerate for a day or two if necessary

BEETS

Children seem to have a reaction to beets ranging from indifference to downright hatred. This phenomenon is strange, because beets are tasty, and all the more curious because beets are by far the sweetest of vegetables. So children either form their opinions from beets that are poorly prepared, stale, or overgrown or from the fact that parents do not seem to like them very much either. Pity. They are fine, honest vegetables when young and very fresh, especially well suited to hearty winter meals. Americans think of beets as fat red things; elsewhere they are called beetroots, to distinguish the root from the highly edible greens, used as a potherb. Swiss chard is the major beet grown only for its foliage, but there are others as well. We would do well to follow the example of Eastern Europeans, who have quite sensibly produced a variety of satisfying soups and stews based on beets, called *borscht*.

Fresh beets are available year round, most plentiful from summer to early fall. For finest flavor and texture, choose those that are small, about the size of golf balls, with good fresh greens attached. Uniform size is best for even cooking. Beets become woody as they grow large and lose sweetness when stale. It is also important that the beets not be trimmed or bruised—if the tapering root end is cut or the stems are trimmed closer than one inch from the root, beets will lose considerable flavor, color, and nutritional value in cooking. However, if they must be stored for a few days, it is advisable to remove the greens before wrapping and refrigerating.

Leaves have so much surface area for evaporation, they will sap moisture (and nutrients) from the root.

Beets are often steamed, baked, or boiled whole until tender, at which point the skins rub off easily. They can then be served simply buttered or flavored with orange, vinegar, cream, or even cheese. Other hot preparations include a variety of soups, red flannel hash, and homemade pasta that is colored a delicate baby pink by the addition of a little beet purée. Beyond pickles and relishes, there are many cold salad possibilities for cooked beets; they team well with herring, potatoes, apples, oranges, mint, horseradish, sour cream, vinaigrette, and salad greens, especially watercress. Beet tops are prepared like other greens, except that the brilliant red stems are usually removed to avoid staining of the leaves. Table salt quickly removes beetroot stains on hands and other surfaces. (Avoid staining fabric, wood, and other porous materials.)

WHAT TO LOOK FOR: small to medium size, uniform roots; fresh greens

WHAT TO AVOID: wilted or rotting greens, trimming, bruises, shriveled skin

HOW TO STORE: refrigerate, wrapped, for a few days

PRIME SEASON: (Indicated by darker shade).

| JAN | FEB | MAR | APR | MAY | JUN | JUL | AUG | SEP | OCT | NOV | DEC |

27

BERRIES

1. Blueberries – uniformly plump and deep in color with the frosty bloom intact. Avoid small green berries – there should not be more than a few per package.

2. Raspberries – dull red, deep red, or sometimes black, dry-looking and slightly fuzzy. Ripe fruit leaves not only the cap but also its center behind – berries should be hollow.

3. Strawberries – uniformly very red with fresh green caps. Bigger is definitely not better, but the large long-stemmed varieties occasionally available are beautiful to serve.

4. Blackberries – shiny, very deep color and no caps. Berries with even one red drupelet will be sour. In fact, commercial berries are likely to be a bit on the sour side, most successful when sweetened.

Whether gathered from the wilds, the garden, the roadside stand, or the supermarket, fresh berries are some of the major glories of summer. Urbanization is taking its toll, but many adults still count among their fondest memories of childhood the annual raid on the wild strawberry patch (its whereabouts carefully secreted from all but the closest friends) and the willing exposure of young hands to the sharp thorns that guarded plump wild blackberries of a sweetness found only in memory. Those of us lucky enough to have indulged in these rites of summer never forget the greedy gorging and the heroic stamina it took to get some of the bounty home to Mother's kitchen. Only the thought of pies, cobblers, shortcakes, and preserves to come encouraged any of the take into the gathering-pail. At a tender age we became expert at selecting only the ripest, sweetest fruit, leaving the rest for another day. Commercial berries are lackluster by comparison. Modern merchandising techniques require firm berries that will travel and keep well. Flavor is the last consideration. The best berries are the local ones acquired right there at the farm stand or, better yet, picked by oneself (some growers allow this). But even supermarket berries are worth trying and are superior to the frozen or canned products.

Summer is berry season, with blueberries available from May through September. Strawberries are now available nearly every day, thanks to imports. In fact, most commercial berries are now available in all but mid-spring and mid-fall. Choose packages of unblemished fruit with no seepage of juice on the bottom. Avoid bruises, mold, or any moisture at all.

Berries should be carefully sorted right after purchase. Discard irregular or bruised fruit (decay spreads rapidly) and refrigerate, wrapped. Rinse immediately before serving, or not at all. Some gastronomes like to rinse berries with a little white wine. Use within a day or two or freeze for later use in cooked preparations.

Fresh, sweet berries are at their best alone or with cream and (if absolutely necessary) sugar. They brighten fruit salads and respond to liqueurs, brandies, eau de vie, and chocolate. There are many traditional berry preparations, including those above plus tarts, jellies, syrups, creams and bombes, mousses, soufflés, muffins and cakes, ice creams and ices. Berries are also turning up in salads along with berry-flavored vinegars – they complement roast breast of duck or chicken, nuts, nut oils, curly endive and other greens, and gently braised leeks.

WHAT TO LOOK FOR:
plump, unblemished berries

WHAT TO AVOID:
bruises, mold, decay, any moisture, especially seepage of juice on the bottom of the package

HOW TO STORE:
sort, then wrap and refrigerate; use as soon as possible

PRIME SEASON: (Indicated by darker shade). This may vary slightly due to local crops.

| JAN | FEB | MAR | APR | MAY | JUN | JUL | AUG | SEP | OCT | NOV | DEC |

BROCCOLI

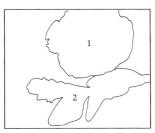

1. Broccoli
2. *Gai lan* (Chinese brocolli)

Cabbages respond well to patient training, managing protean feats that few plants have accomplished. Broccoli, named for its Italian forebears, is but one of cabbage's incarnations–thick flower stalks with masses of tight buds that are harvested and eaten before they open. If you think this is the same lesson learned by cauliflower, you are quite right. In fact, the two are nearly identical botanically and have always been a source of confusion for those who like their classifications tidy. Purple cauliflower looks suspiciously like broccoli, and there are broccolis with a decided purple cast and those, called 'heading' broccoli, that defy distinction from the white sibling. And there is also the Chinese cousin *gai lan* that is often called broccoli, though it is really just another of the many relatives in the cabbage family. All have a healthy vitamin payload and a fresh green crunch, so long as they can elude cooks who insist upon overcooking vegetables.

Broccoli and *gai lan* are available year round, at their best and most plentiful without the heat of summer. Broccoli must have tight green heads with no yellowish sprouting to the flowers, while *gai lan* can show pretty little white blooms. Both must be crisp, with fresh leaves. Wilting, rust, and yellowing will always result in disappointing flavor. They can be wrapped and refrigerated for a few days if necessary.

Broccoli is usually trimmed (often the stems are peeled) and steamed or blanched until crisp/tender, then served with butter, lemon butter, or a butter sauce such as *hollandaise* or *maltaise*. Rich white sauces, perhaps flavored with cheese, are also usual, as is purée or cream soup. Olive oil, onions, and garlic are good partners. Broccoli makes a fine *crudité*, especially if we cheat and blanch it for a few seconds. Once blanched it is ready for a wide variety of salads and stir-fried dishes. For salad it is best to dress broccoli at serving time, for an acid marinade will yellow it. Broccoli is very much at home with fish, eggs, turkey, ham, and beef. It is also delicious in Asian-style noodle dishes, hot or cold, flavored with sesame or peanut butter, garlic, and chiles.

WHAT TO LOOK FOR:
fresh greenness, tightly headed broccoli

WHAT TO AVOID:
wilt, rust, yellowing, loose or blooming heads on broccoli

HOW TO STORE:
refrigerate, wrapped, for a few days

PRIME SEASON: (Indicated by darker shade).

| JAN | FEB | MAR | APR | MAY | JUN | JUL | AUG | SEP | OCT | NOV | DEC |

BRUSSELS SPROUTS

There is good news for those who have sworn off the lowly sprout because of memories of mushy, malodorous messes that sullied plate, palate, and the air in general. The truth is that Brussels sprouts are fine-flavored, even sprightly, so long as the traditional Anglo-American cooking method–extreme overcooking–is avoided. As cabbages go, this one is quite delicate when young and fresh and properly undercooked so that it retains its good green color, its crisp center, and its vitamins and minerals. Children of all ages cannot fail to delight in the sprout's unique growth pattern, the fat stalk with spiraling rows of frosty little blue-green buds (*Rosenkohl*, 'rose cabbage' in German).

This is a cold-weather crop, now available year round, least available in summer, with the best sprouts to be had from October into February. Mid-fall is the best time to find fresh whole stalks, the fun form, while the familiar one-pint paper tubs are the rule for the rest of the year. Choose bright green sprouts that are firm and compact and of the smallest size available–one or one and one-half inches in diameter is usual. Avoid yellowing, wilt, flabbiness, rot, and any raggedness that might be caused by insect damage. Remove any irregular outer leaves and refrigerate, wrapped, for a day or two if necessary. The fresher the sprouts, the finer the flavor.

Depending upon size, seven minutes of steaming or simmering is about the maximum for good color, texture, and taste. Many good cooks, after trimming the stem end, pierce a cross deep into the base to promote even, rapid cooking. It is a valuable technique. It is also wise to select for uniform size and then trim sprouts as necessary. Brussels sprouts team well with white or butter sauces, tomatoes, cream, sour cream, cheeses, other vegetables, and nuts, especially almonds, pecans, walnuts, and chestnuts. Water chestnuts are a nice crunchy complement. There are also many salad possibilities, as well as purées, casseroles, and pickles. Sprouts are especially good with red meats, game, and poultry.

WHAT TO LOOK FOR:
fresh greenness, even size

WHAT TO AVOID:
yellowing, wilt, flabbiness, rot, insect damage

HOW TO STORE:
refrigerate wrapped, unwashed, for a day or two

PRIME SEASON: (Indicated by darker shade).

| JAN | FEB | MAR | APR | MAY | JUN | JUL | AUG | SEP | OCT | NOV | DEC |

CABBAGE

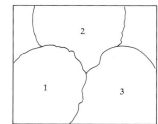

1. Savoy
2. Green
3. Red

The lowly cabbage just got a new press agent: the modern chef who treats it with respect. Now cabbage leaves are enveloping all manner of costly delicacies with surprisingly dainty results; surprising, that is, for all who associate cabbage with humble kitchens and reeking hallways. As families go, this one is large indeed, encompassing not just head cabbages but also kale, collards, kohlrabi, rape, turnips, mustard greens, broccoli, cauliflower, Brussels sprouts, radishes, and choy. They have one familial characteristic that transcends variety of appearance, texture, and taste—they all deliver fresh, delicate flavor when they are gently treated. Stench and mushy wateriness develop only with overcooking. So, we can only hope that these kitchen staples will receive even more justice and continue to bring their savory goodness and considerable nutrition to generations to come.

Head cabbages are in almost uniform supply throughout the year, though there is a bit less quantity and quality in summer. Whether green, the crinkly green called savoy, or red, the criteria are the same—heads should be solid, heavy, and fresh-looking. Avoid wilt, flabbiness, and rust. Fresh cabbages have the most delicate flavor, and they are likely to have a fair number of their dark, loose outer leaves attached. Growers trim these leaves as they wilt, so a cabbage that is well trimmed down to its pale inner layers is likely to have some age on it—still acceptable but not at its best. Cabbages can be wrapped and refrigerated for a week or so if very fresh.

Cabbages are often shredded for coleslaw and other salads, where they team well with apples, carrots, onions, oranges, raisins, nuts, and even pineapple. Sauerkraut is the famous pickled cabbage, usually served with cured or smoked pork products, goose, or duck. Another famous dish is plain boiled cabbage served with corned beef or smoked pork butt. Red cabbage is especially good braised with loin of pork and chestnuts. Green cabbage leaves are often stuffed with meat and rice and are simmered in a rich tomato sauce. There are also many possibilities for soups. Newer dishes emphasize the delicate flavor of lightly cooked cabbage teamed with the likes of salmon, partridge, and even berries.

WHAT TO LOOK FOR:
solid, heavy, fresh-looking

WHAT TO AVOID:
wilt, flabbiness, rust, excessive trimming

HOW TO STORE:
refrigerate loosely wrapped, for a week or so

PRIME SEASON: (Indicated by darker shade).

| JAN | FEB | MAR | APR | MAY | JUN | JUL | AUG | SEP | OCT | NOV | DEC |

CHINESE CABBAGE

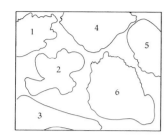

1. Choy sum
2. Shanghai bok choy
3. Long bok choy
4. Gai choy
   (Chinese mustard greens)
5. Short bok choy
6. Tientsin bok choy
   (Chinese cabbage or napa)

In China, the choy of cooking are plentiful and varied, and they happen to be cabbages, though they display the same clever diversity as their Western cousins–broccoli, cauliflower, kale, collards, turnips, and the like. Some choy are prized for their thick, crunchy stems of white or green, the leaves used sparingly, while others are leafier and used whole. Long bok choy and Chinese cabbage (napa or Tientsin bok choy) are the only two that have gained any sizable following in the Western kitchen, abetted by the California/health food lifestyle. All deserve more attention and accessibility beyond the Chinese markets in large cities. Among their virtues are considerable nutrition, year-round availability, and a wonderfully fresh flavor and crunch.

Choy need warm to cool growing conditions; they are always being harvested somewhere. Markets that carry a good selection will have most of them at any given time, with less availability in midsummer. Look for fresh greens and plump stems. Avoid wilt, rot, and rust. Choy are at their best when very fresh, but they can be wrapped and refrigerated for a few days if necessary.

Chinese mustard greens–gai choy–are most often salted and left to pickle for a few days; then they are rinsed, sliced, and served for their own sake or stir-fried with meats or seafood, especially squid. When used fresh they are often blanched before the final cooking, as is Shanghai bok choy. The others are enjoyed mostly for their crunchy stems, so these choy are always trimmed into even-sized pieces and quickly cooked to preserve their texture and flavor. Most of the leaf is either discarded or added at the very end of the cooking process so that it will not be overdone. Western cooks should feel free to experiment with salad and soup uses for the good green leaves that otherwise might go to waste. Choy are most often used in soups, dumplings, or in stir-fried dishes where they complement the tastes and textures of many foods–pork, chicken, beef, lamb, seafood, and other vegetables. The usual flavorings are soy sauce, garlic, ginger, chiles, rice wine, sesame oil, and a wide variety of spices and condiments.

WHAT TO LOOK FOR:
fresh greens and plump stems

WHAT TO AVOID:
wilt, rot, and rust

HOW TO STORE:
refrigerate wrapped in moist paper, for a few days if necessary

PRIME SEASON: (Indicated by darker shade).

| JAN | FEB | MAR | APR | MAY | JUN | JUL | AUG | SEP | OCT | NOV | DEC |

CARDOONS

One of the stranger vegetables, one that has never really caught on beyond the Mediterranean, is a close relative of artichokes. But here, the leaf stalks are eaten and not the flower buds. Cardoons are winter vegetables that finish their lives in some sort of blanching operation, probably wrapped in burlap or paper these days, to whiten them and mellow the intense flavor that sunlight would produce. They really do taste a bit like artichokes, though there is a celerylike quality that echoes the physical similarity.

Cardoons are less than plentiful in northern Europe and are extremely rare in the U.S. Markets that cater to Italian or Provençal tastes are the best bet, and the season is limited to midwinter, especially Christmastime. A bunch might be anywhere from one and one-half to four or five feet long, but is usually stripped of its outer ribs and trimmed to the smaller dimensions. Choose bunches that are as crisp and fresh as possible. Avoid wilt and excessive rust. Cardoons may be wrapped and refrigerated for a few days if necessary.

Very white, tender inner ribs can be eaten raw, as they are in the Piedmont region of Italy, as a carrier for the wonderful 'hot bath' of olive oil, butter, anchovies, and garlic called *bagna cauda*. They are always trimmed to remove the skin and heavy fibers. The leaves are discarded. Except for in *bagna cauda*, cardoons are simmered until tender, then used in a variety of preparations; they may be fried, braised, creamed, gratinéed, or used in soup or salad. Like artichokes, cardoons should be trimmed with a stainless steel knife and immediately acidulated with lemon juice or vinegar to prevent browning. Home gardeners who decide to grow cardoons may want to let a few of them flower, dry the flowers, and use them to curdle milk for making fresh cheese.

WHAT TO LOOK FOR:
fresh, crisp bunches, not too large

WHAT TO AVOID:
wilt, excessive rust, rot

HOW TO STORE:
wrap and refrigerate for a few days if necessary

PRIME SEASON: (Indicated by darker shade).

| JAN | FEB | MAR | APR | MAY | JUN | JUL | AUG | SEP | OCT | NOV | DEC |

CARROTS

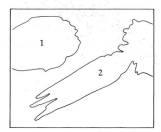

1. Baby (finger) carrots
2. Bunch carrots

Fifty million rabbits can't be wrong. Carrots are delicious and nutritious and even generations of motherly entreaties involving eyesight and curly hair have not managed to diminish their popularity. Most adults like them and even youngsters may respond when carrots are young and fresh and their considerable natural sweetness is enhanced by honey or brown sugar and sweet spice. And inquisitive children might be pleased to learn that they were *right*–the weed that abounds in roadsides and previously cultivated areas now gone to seed, the one that adults call Queen Anne's lace, *is* wild carrot, probably a fugitive from the garden centuries ago.

Fresh carrots are available year round in several forms. The best are the medium-sized ones sold in bunches with their tops (the greens are edible in soups and salads but are rarely used). The condition of the tops tells how fresh they are, and these carrots are likely to be of a size and type that is finest in flavor and tenderness. The familiar one-pound plastic bags of trimmed carrots are of questionable freshness, as are the mammoth ones sold loose. Tiny, or finger carrots are very popular in Europe and are now being grown in the U.S. as well. They may be mature dwarf carrots or immature average ones, and they can be very tender and

sweet. Choose plump, crisp carrots that are smooth and shiny with very fresh greens, if any. Avoid wilt, shrivel, greenish coloring, rot, and any trimmed carrots that are sprouting. Fresh carrots can be wrapped and refrigerated for a week or so if necessary. Be sure to trim the tops from bunch carrots as soon as you get them home, or the greens will rob the root of moisture and nutrients.

Carrots are versatile players in both starring and supporting roles. If possible it is best to cook them whole and unpeeled to preserve their flavor and nutritional value. The skins will then rub off easily. In fact, very young carrots need not be peeled at all. Remember that huge carrots are not only less tasty, but the woody core will probably have to be removed. Steamed, boiled, baked, stewed, stir-fried, or puréed, carrots can accompany any meat or poultry. Common additives are butter, orange, nutmeg, ginger, parsley, dill, and other herbs and spices, cream, and sweeteners. Carrots also mix well with other vegetables, though the peas and carrots medley of childhood is one of the least interesting. Carrots are fine in soups and are indispensible in the stockpot. Raw, carrots are often used for *crudité* and for salads, where they go well with cabbage and raisins. And carrot cake lovers know that the flavor is delicious in baking.

WHAT TO LOOK FOR:
plump, crisp, smooth and shiny roots, fresh greens if any

WHAT TO AVOID:
wilt, shrivel, greenish coloring, rot, sprouting

HOW TO STORE:
wrap and refrigerate for a week or two if necessary

PRIME SEASON: (Indicated by darker shade).

| JAN | FEB | MAR | APR | MAY | JUN | JUL | AUG | SEP | OCT | NOV | DEC |

CAULIFLOWER

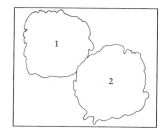

1. Purple
2. White

Cauliflower is hardly the most glamorous of vegetables but, like its siblings, the cabbages, it has been much enjoyed through the ages. The ancient Romans doted on it, and then it fell victim to the barbarian scourge, disappearing for centuries. The Renaissance brought a reflowering of many things, the cabbage flower but one. And if we needed evidence that it finally arrived, a bit of culinary flattery would do it – an eighteenth-century French chef created a dish, probably a purée, that would forever link cauliflower with Louis XV's Madame Dubarry.

Mark Twain called it 'nothing but cabbage with a college education,' and he was reasonably accurate. Cauliflower is simply a cabbage that has been trained to produce firm bunches of flowers, and some modern varieties have even been educated to shelter the curd (the technical name) from sunlight by wrapping the leaves around it. Less precocious varieties have to be tied to effect this necessary blanching that produces a pale, delicately flavored result.

Most of the cauliflower on the market is white or ivory, and it should be solid, heavy, and unblemished by rust. Look for fresh green leaves. Avoid spreading or flabbiness, dark spots, and yellowed or wilted leaves. Purple or green varieties can be found occasionally, and they should be judged similarly. Refrigerate, if necessary, for a day or two. Cauliflower is a cold weather vegetable, at its best in fall and winter, but the standard white variety is available year round with smaller, usually less desirable crops in summer.

Sauce Mornay, a white sauce flavored with Gruyère or Parmesan cheese, is perhaps the most famous of cauliflower partners, though the vegeta-

WHAT TO LOOK FOR: solid, heavy, unblemished curd, fresh green leaves

WHAT TO AVOID: spreading, flabbiness, rust, wilted or yellowed leaves

HOW TO STORE: refrigerate wrapped, for a day or two

ble teams well with other cheeses, lemon butter, and other rich concoctions. It can also be puréed, curried, pickled, deep-fried, served raw as a *crudité*, or dressed with a cold sauce for salad. It is essential that cauliflower be briefly cooked – under five minutes of simmering should be plenty for cut florets, a more successful method than cooking the whole curd. The addition of a teaspoon or two of lemon juice to the cooking water may help to preserve whiteness. Overcooking leads to mushiness and noxious taste and smell, not to mention nutritional loss. Cauliflower greens and stem are highly edible, especially good in soups and stir-fries.

PRIME SEASON: (Indicated by darker shade).

| JAN | FEB | MAR | APR | MAY | JUN | JUL | AUG | SEP | OCT | NOV | DEC |

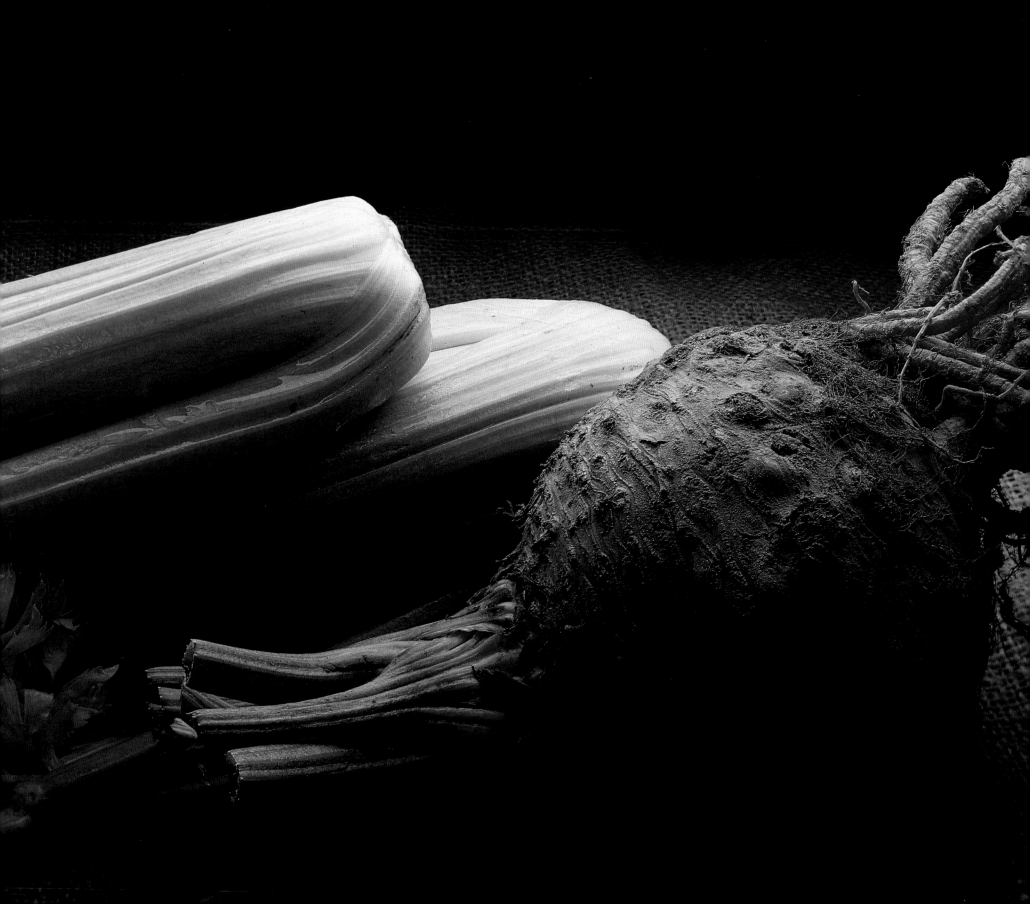

CELERY

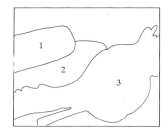

1. Celery heart
2. Celery
3. Celery root (celeriac)

Anyone who has ever wondered why Bloody Marys often come with a celery stalk as a swizzle stick may be cheered to learn that celery has an ancient Roman tradition as hangover preventive and remedy. Only Old Mr. Boston knows for sure where this modern tradition started, but it is colorful, tasty, and it couldn't hurt. The Romans cultivated celery and may have eaten the wild variety as well, or just used it for medicinal and flavoring purposes, but the wild one is intensely flavored, not well suited to eating raw or in any great quantity. Sometime in the sixteenth century the taming process began anew, and celery eventually became the pleasant, fat, crunchy vegetable it is today. While domestication was under way another variety was tamed as well, this one for its roots, called celeriac, celery root, or knob celery. Only the crisp root with its assertive celery flavor is eaten, the greens discarded.

Both celery and celeriac are available year round, though celeriac is in very short supply in summer. Most of the celery on the market is of the greenish sort called Pascal, and it can be had in large 'stalks' (individual stems are called ribs) or smaller packages of trimmed, usually rather white celery called celery hearts. Both should be very crisp with good fresh leaves, if any, and no evidence of wilt or rot.

Very green celery is intensely flavored and most suitable for cooking, while the fat whiter varieties, often blanched by banking them with soil, are most pleasant for eating raw. Either should keep well for several days if kept moist, wrapped closely, and refrigerated. Celeriac is less likely to be woody if it is under four inches in diameter. Always ugly and warty, it should, however, be plump and heavy, so avoid shriveling and softness. It too keeps well, wrapped and refrigerated, but the greens, if any, should be removed before storage.

Celery is very popular in salads and as *crudité* with a wide variety of dressings, dipping sauces, or stuffings, often including cheese, peanut butter, or olives. Indispensible in the stockpot, it also turns up in soups, stews, casseroles, and stuffings. Celery participates in a variety of vegetable combinations and stir-fries and is sometimes featured on its own, often braised. Celeriac is most often peeled, cut into fine *julienne,* and marinated in a mustard mayonnaise to be served forth as the famous hors d'oeuvre called *céleri rémoulade.* Celeriac is also a fine cooked vegetable in buttered *juliennes,* for flavoring soups and stews, or as a purée–the flavor is sometimes gentled by the addition of potato–to accompany roast meats and poultry.

WHAT TO LOOK FOR: crisp ribs, fresh leaves; root should be plump, hard, and heavy

WHAT TO AVOID: wilt, rot; shriveled, soft, or overgrown roots

HOW TO STORE: refrigerate wrapped, for a few days

PRIME SEASON: (Indicated by darker shade).

	JAN	FEB	MAR	APR	MAY	JUN	JUL	AUG	SEP	OCT	NOV	DEC
Celery												
Celery root												

CHERIMOYAS

All fruits have armor of some sort, the better to protect the precious seeds they carry. But cherimoyas look as if they were designed by a pangolin or some other unlikely beast. The exterior is truly strange, but the innards are luscious. Cherimoyas are native to mountainous tropics of the New World, and tropical is just the way they taste. The pulp is sweet and juicy with a flavor more like litchis (another well-armored fruit) than anything else. You might find them marketed under the names custard apple or sweetsop, though there are other fruits more properly entitled to those names.

Cherimoyas are mostly winter fruits, though there are sometimes a few in early fall as well. They come in a variety of shapes and sizes. Choose fruit with dull khaki-green skin that is just beginning to soften. For added sweetness, you may age the fruit at room temperature just until it begins to brown a bit, but take care not to let it go too far–cherimoyas are perishable. Do not buy one that is already brown. Ripe cherimoyas should be used promptly, but they can be refrigerated for a day or two if necessary.

Cherimoyas are a little tricky to eat. They are fibrous and riddled with fairly large shiny black seeds. A spoon is probably the best tool; it can be used to scrape the custardy goodness from the fiber, but the seeds will come too. Just spit them out or follow a more genteel method of disposal. Though they are most often eaten straight, there is no reason why cherimoyas could not be strained and the purée used for a variety of desserts–custards, tarts, bombes, mousses, creams, ices, and ice cream.

WHAT TO LOOK FOR: plump, dull khaki-green, slightly soft

WHAT TO AVOID: brown, bruised, mushy fruit

HOW TO STORE: age for a few days at room temperature if desired, then use promptly or refrigerate for a day or two

PRIME SEASON: (Indicated by darker shade).

| JAN | FEB | MAR | APR | MAY | JUN | JUL | AUG | SEP | OCT | NOV | DEC |

CHERRIES

These early sweet cherries have not achieved the richness of color and flavor of the main crop, but are still acceptable to those who cannot wait. Be sure to handpick the darkest ones.

Would that life *were* just a bowl of cherries. Sweet, juicy, and delicious–that's what cherries are, unless they are tart, juicy, and delicious, in which case they are better with a little experience in the kitchen. There are actually three kinds of fresh cherries–sweet, sour, and sweet-and-sour hybrids. Sweet cherries are the most plentiful, grown most extensively in the U.S., Romania, Germany, and Italy. They range in color from yellow to mottled yellow-and-pink to red to black. Sour cherries and the hybrids are usually smaller, also colorful, and versatile for cooking and for spirit making. Kirsch, also called kirschwasser, is a sublime eau de vie, very popular for sipping and in desserts, and maraschino and ratafia are other delicious cherry-flavored spirits. There are even some wild cherries to be found, ranging from inedible to quite good when cooked and sweetened. Next to man, birds are the major consumers of cherries, showing that they have good taste in summer fruit.

Good sweet cherries come to market first in May and are at their best in June, July, and August. Sour and hybrid cherries start a little later, in late June. Some Southern Hemisphere crops do come to northern markets in midwinter, but they are likely to be pallid and not very well flavored due to premature harvesting. For best results, handpick plump, glossy cherries with stems attached that are the darkest in color of that variety. Avoid mushiness, rot, dull or shriveled skin, splits or cracks. Cherries do not improve much after picking and are perishable, so refrigerate them in loose wrapping and enjoy them within a day or two.

Sweet cherries–Bing, Royal Ann, Vignola, Early Burlat, and such–are most often eaten out of hand. They also combine with other fruits for salads. They are sometimes used in cooking, but sour cherries and hybrids–Morellos, Amarelles, Dukes–have more zing. Traditional preparations featuring cherries include Black Forest torte, duckling Montmorency, and the homey flanlike dessert from Limousin called *clafouti*. Cherries are also used for jams, jellies, and preserves. There is a plierslike appliance for efficiently pitting cherries, but home cooks who do this sort of thing often claim to be very quick with a hairpin or a paper clip.

WHAT TO LOOK FOR:
plump, glossy cherries with stems, deep color

WHAT TO AVOID:
mushiness, rot, dull or shriveled skin, cracks or splits, moisture

HOW TO STORE:
loosely wrap and refrigerate for a day or two

PRIME SEASON: (Indicated by darker shade).

| JAN | FEB | MAR | APR | MAY | JUN | JUL | AUG | SEP | OCT | NOV | DEC |

CHICORY

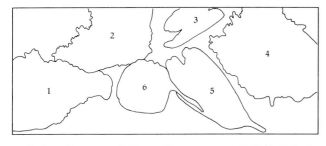

1. Curly endive
2. Escarole
3. Belgian endive
4. Italian curly endive
5. *Radicchio di Treviso*
6. *Radicchio di Verona*

Who is chicory, what is she? Why, it depends upon where you live. Science has decided that all the chicories and endives are botanically *Chicorium*, but popular usage has confused the issue greatly. Names vary not just from country to country, but regionally as well. The cold facts are these: the chicories are green chicory, Belgian endive, and red chicory; the endives are escarole and curly endive. But things are not that easy at the market, so know the plant and do not expect clarity from the greengrocer, wherever he may be. Once past the obstacles of nomenclature, people find much to enjoy in these vegetables, whether they are served raw or cooked. All have an appealing bitterness that ranges from quite delicate in the pale varieties to rather pungent in the deeply colored ones.

GREEN CHICORY
This has been a popular salad vegetable and potherb for centuries. Some varieties are grown exclusively for their fat taproots, which are dried and toasted for use as a coffee additive. The greens are common in Europe but rarely seen in the U.S. except for the one called Catalonian chicory or Catalonia, which is actually marketed as dandelion during the winter months. Green chicories can be judged and prepared in the same ways as their near look-alike.

BELGIAN ENDIVE
Called *witloof* (Flemish), this is a mid-nineteenth-century accident that came as a happy surprise to a gardener who had intended to grow winter greens in a greenhouse. Another version of the story involves a winter heat wave that caused sprouting from roots that were stored in a dark cellar. However it happened, it was good. So controlled conditions were developed for forcing this pale little spike that is so popular (and so costly) in Europe and, increasingly, in the U.S. It is grown almost exclusively in Belgium and if it were not grown in darkness, it would be green chicory. It is at its best in cold weather, in smaller supply or absent in midsummer.

Belgian endive should be tightly headed and creamy white blending to pale yellowish at the tips of the leaves. Avoid any browning or greening. Many greengrocers fail to realize that it must be stored within the protective darkness of its blue paper packaging, for it will most assuredly begin to spread and green with exposure to light. Once it begins to green it loses its delicate flavor and is no longer worthy of its fancy price tag. If very fresh, it should keep well for a day or two, wrapped in opaque paper and refrigerated. Belgian endive is excellent in the salad bowl or as an hors d'oeuvre, especially with bleu cheese, walnuts, and pears. In France and Belgium it is often braised or wrapped with ham and served au gratin. For cooking or for serving raw, be sure to remove the bitter cone-shaped base.

ITALIAN RED CHICORIES
The most widely grown are the round-headed one called *radicchio di Verona* and the leafier one called *radicchio di Treviso*. They should be fresh and crisp

PRIME SEASON: (Indicated by darker shade).

| JAN | FEB | MAR | APR | MAY | JUN | JUL | AUG | SEP | OCT | NOV | DEC |

and are not very interesting, or economical, once wilt has set in. They can be wrapped and refrigerated for a few days if necessary. The appealing bitterness of *radicchio* is much prized in salads. Though they can be cooked, red chicories will turn an unpleasant rusty brown when overheated and are usually more successful raw.

TRUE ENDIVE

It is represented on the table by escarole, the one with the fat, fleshy leaves, and by several curly endives that are often called chicory or *chicorée frisée.* To add to the confusion, there is the marketing term *endive-escarole*, which is usually applied to escarole but might be applied to curly endive as well. And escarole is also known as Batavia or Batavian endive in England and Batavia or just *chicorée* in France. Escarole and curly endive are available in fairly steady supply throughout the year, though the quality is usually rather poor in summer. They should always be very fresh and crisp, and the smaller, paler heads have the most delicate flavor. In Europe, curly endive is often blanched, but the American, French, or Italian variety is almost always reserved for the salad bowl, while escarole is also cooked in a variety of preparations–braised, creamed, sautéed, souffléd, in soup. For salad, both are good with crisp bacon and other flavorful ingredients.

WHAT TO LOOK FOR:
fresh, crisp greens, pale *chicons* of Belgian endive (chicory)

WHAT TO AVOID:
wilt, rot, browning, greening of Belgian

HOW TO STORE:
refrigerate wrapped for a day or two

51

CHIVES

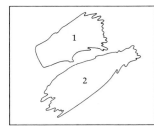

1. Potted chives
2. Chinese chives

These potted chives are in a typical root-bound condition and need to be repotted if they are to be kept growing. There are two varieties of Chinese chives: the green ones shown here, and very pale ones that have more delicate flavor.

If anyone needed to be convinced of the palatability of onions, he would do well to try the most delicate member of the clan, chives. They even look gentle and inviting, one of the reasons that people like to grow them on sunny windowsills. The truth is that no sane person in the world quarrels with the wholesomeness or savor of the onion family (garlic is the loud exception), but too few of us get to enjoy fresh chives, for the soft springtime flavor is echoed in fragile shelf life, the kiss of death so far as produce growers, shippers, and marketers are concerned. While there are small local crops sold in fancy markets at fancy prices, the sunny windowsill routine is a more reliable source. The green-thumbed among us can keep a little pot of chives going from one spring to the next. Variations that are more robust in flavor and texture are the so-called Chinese chives that are so abundant, alas, only in Asian markets.

Fresh chives are available year round but always in short supply. Cut chives sold in little bunches must be very fresh, so avoid wilt, rot, and moisture.

They do not keep well, but can be wrapped in damp paper and refrigerated for a day if necessary. The Chinese chives frozen whole will keep their flavor for a month or two. Some markets carry little pots of growing chives that can be sustained if they are repotted, harvested randomly, and given plenty of sunlight.

Chives are rarely cooked (except in omelets), but are sprinkled onto a finished dish at the last moment. They are either snipped with scissors or carefully sliced with a very sharp knife using a sawing motion to avoid bruising. This mincing is always done just at serving time, for the delicate flavor goes off faster than with any other onion. Chives are especially good with eggs, potatoes, fish, creamy soups, and many salads. Because of their color and mildness, they often flavor 'white food,' including sour cream and cream cheese. Chives will usually be wasted on a highly seasoned dish, or any preparation in which the dominant flavor is onion or garlic. Chinese chives are most often used with noodles, eggs, and stir-fried dishes.

WHAT TO LOOK FOR:
freshness, good green color

WHAT TO AVOID:
wilt, rot, moisture, yellowing

HOW TO STORE:
refrigerate wrapped in damp paper, for a day or two, or freeze whole for two months if necessary

PRIME SEASON: (Indicated by darker shade).

| JAN | FEB | MAR | APR | MAY | JUN | JUL | AUG | SEP | OCT | NOV | DEC |

COCONUTS

As tropical fruits go, the coconut surely looks as exotic as they come. A big fat thing that resembles dead wood in its whole state, a small hairy brown bowling ball when husked, it would hardly inspire confidence in those unaware of the treasure within. For people in temperate zones it is a pleasant dessert fruit that is remarkably easy to come by, owing to its outstanding shelf life. Yet many people prefer processed products to wrestling with the tough shell. The coconut is truly a hard nut (actually a seed) to crack. Citizens of the tropical Pacific and the West Indies view the coconut palm with more respect, for it is a life force more surely than the date palm is to desert dwellers. The coconut provides all manner of foodstuffs from milk to sugar to oil, building and household materials, and is the basis for whole economies. The treat that never gets beyond the tropics (perishability is the problem) is the green coconut, easier to crack than the mature one and possessed of soft jellied flesh and sweet milk.

Coconuts are in the market all year, most plentiful in fall and early winter. Fresh coconuts have solid shells with the three 'eyes' intact and firm. Avoid cracks, holes in the eyes, mold, or moisture. They should feel heavy and the coconut water inside should slosh around audibly. Fresh coconuts will keep for at least a month at room temperature, and longer refrigerated.

Placing coconuts in a moderate oven for twenty to thirty minutes should crack the shell. If not, a hammer will do the trick. First pierce the eyes with a screwdriver or ice pick and drain the water, reserving it for sauces, beverages, or for moistening cakes. Pry the meat from the cracked shell, pare away the brown skin, and grate the flesh. It is then ready to sprinkle onto fruit salads (ambrosia), or for cakes, pies, cookies, custards, and puddings. Shredded coconut is sometimes toasted. Chocolate is a popular partner. The flavor of coconut is often extracted by hot water, called coconut milk if you start with equal parts water and shredded meat, or coconut cream if you get a more concentrated brew using less water. Coconut milk and cream are used for various desserts and drinks, and in curries and sauces for Indonesian satay.

WHAT TO LOOK FOR:
solid, heavy coconuts with water that sloshes around

WHAT TO AVOID:
cracks, holes, mold, moisture, lightweight or empty nuts

HOW TO STORE:
room temperature for about a month or refrigerate longer

PRIME SEASON: (Indicated by darker shade).

JAN	FEB	MAR	APR	MAY	JUN	JUL	AUG	SEP	OCT	NOV	DEC

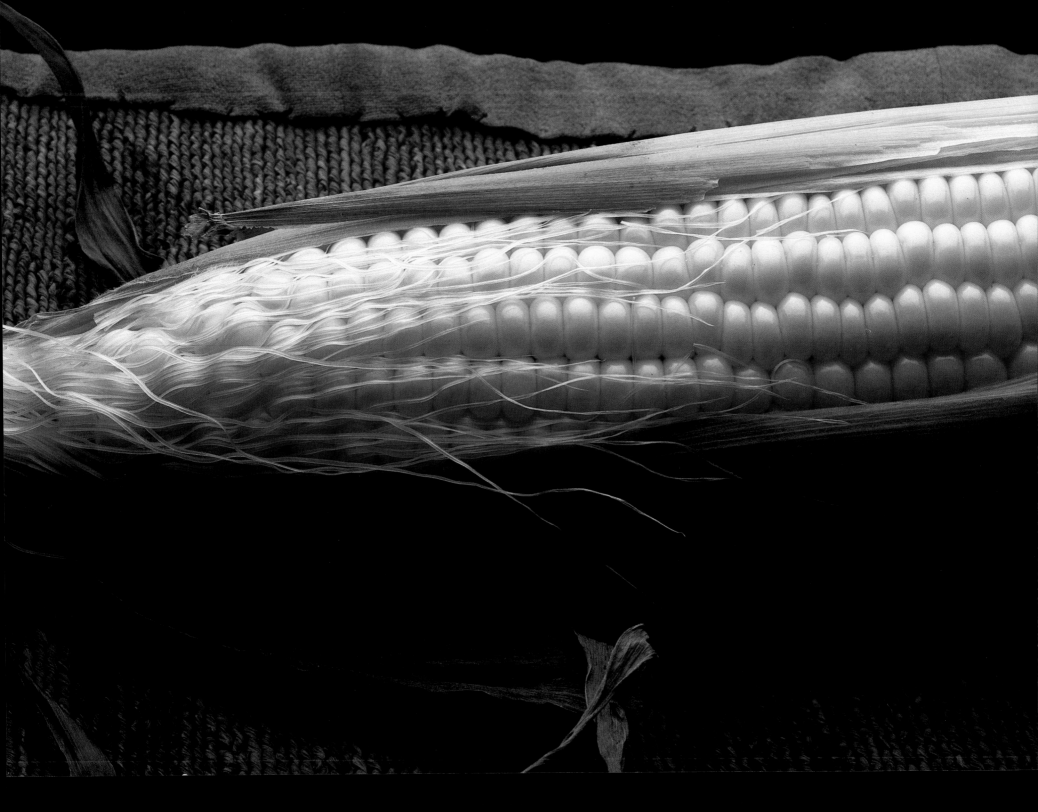

CORN

Corn is a European term that simply means kernel, and was understood to indicate the grain most plentiful and popular. Nowadays, it usually means maize, a variation on grass that is perhaps the most important of all native American plants. It is a remarkably adaptable, hardy, and high-yielding thing, now grown in any area that gets some warm weather. As plants have so often played a major part in the history of mankind, this one is so nutritious, durable, and portable when dried that it is responsible, in part, for the success of Mayan, Aztec, Inca, and North American Indian civilizations, European colonization in the New World, westward pioneering, the African slave trade, and famine relief in many countries of the world. Only the feisty Irish refused it during the potato blight of 1846. Of all the cereals of the world, only rice is grown in larger quantities, and though corn is less visible on the table than many other vegetables, it is hiding in beef, veal, pork, poultry, and dairy products, so that many diets never see a day without it. And, of course, movies could never be the same without the popped variety.

Sweet or green corn, the fresh stuff, is available year round, always growing somewhere. Unfortunately, it is only at its best when cooked and eaten within a few hours, or better still, a few seconds of picking. Sugars convert to starches and no modern ploy can halt the process. If you cannot grow your own or get to a farm stand at harvest time then you will have to content yourself with the market stuff, which can be acceptable to all but the most hardline fanciers. Look for fresh green husks, fresh-looking silk (the external ends are always dry), and cobs that are filled with plump, even rows of kernels. Avoid dry or yellowed husks, shriveled, irregular, or undeveloped kernels, and rot or worm blight. Most of the corn on the market is yellow, though white is popular in some areas, especially the American South and Italy. Mixed varieties are grown for local sale. Use as soon as possible. If it must be stored, leave whole and refrigerate.

Fresh sweet corn seems to be especially at home with its fellow New World natives–tomatoes, lima and other beans, peppers, squash, lobster, sweet potatoes–and with naturalized citizens such as okra and greens. It goes well with the flesh of any of the beasts that it helps to fatten, especially pork, and with cream and cheese. Purists insist that it be quickly boiled, steamed, or roasted and served on the cob with only butter, salt, and pepper.

WHAT TO LOOK FOR:
fresh green husks; plump, even rows of kernels

WHAT TO AVOID:
dry or yellowed husks; shriveled, irregular, or underdeveloped kernels; rot or worm blight

HOW TO STORE:
refrigerate in the husk for a day or two

PRIME SEASON: (Indicated by darker shade). This may vary slightly due to local crops.

| JAN | FEB | MAR | APR | MAY | JUN | JUL | AUG | SEP | OCT | NOV | DEC |

CRANBERRIES

Long before any Europeans paid them a visit, American Indians of the Northeast were grinding cranberries with bear meat or venison to make pemmican, the traditional dried concoction that sustained the tribes and has fascinated schoolboys and Boy Scouts ever since. And the all-American favorite, cranberry sauce, is one of those Indian inventions so generously shared with the newcomers. It is comforting to think of a relish of crushed cranberries, lightly sweetened with maple sugar or honey, gracing the tables of the first Thanksgiving. It is a handsome, shiny red berry with a sprightly tartness, and it has become as American as Mom's apple pie, which might very well be Mom's cranberry-apple pie.

Cranberries are harvested in the fall starting in September and are available fresh through December, usually in the familiar twelve-ounce plastic bags from Ocean Spray (which has definitely cornered the market). Their high acid content makes them by far the most durable of berries, among the most durable of fruits. Choose plump, hard, shiny berries that are as red as possible. Avoid bruising,

rot, and moisture. Store refrigerated until needed, then sort out and discard soft berries and any tiny stems. For longer keeping, pop the whole bag into the freezer and store for a month or two, even longer at sub-zero temperatures. Freezer damage to flavor and texture is slight.

All too often cranberries are thought of as an obligatory addition to the holiday table, and then in a form that tastes more like the can they came in than the berries themselves. Fortunately, there are many traditional uses for fresh cranberries, including homemade relishes, pies, cakes, breads, steamed puddings, jellies and preserves of all sorts, and juice. Practitioners of the New American Cooking (as well as nouvelle cuisine chefs) are celebrating the virtues of this native with innovative preparations of poultry, game, pork, even *charcuterie* (pâtés, terrines, sausages, and so on), and desserts (how about cranberry soufflé?). The refreshing tart flavor of fresh cranberries also marries well with apples (as mentioned) and apple juice, nuts, citrus fruits, and brandy and other spirits.

WHAT TO LOOK FOR:
red, plump, hard, shiny

WHAT TO AVOID:
bruising, softness, rot, moisture

HOW TO STORE:
refrigerate as is for a week or so, freeze for two months

PRIME SEASON: (Indicated by darker shade).

| JAN | FEB | MAR | APR | MAY | JUN | JUL | AUG | SEP | OCT | NOV | DEC |

Cucumbers

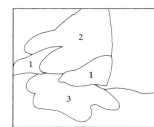

1. Seedless
2. Standard
3. Kirby

Are they really cool? Well, not miraculously so, but when they are growing they are considerably cooler than, say, tomatoes. And after they are picked, cucumbers do have a remarkable way of taking the swelter out of summer, for those who eat them and for those who apply them to the skin. Cucumbers also help to put out the fire when spicy food is the heat source. Coolness is about all there is–no powerhouse of nutrition here–but cucumbers have been a great favorite since ancient times.

Cucumbers come in a variety of sizes, from tiny gherkins to the huge seedless varieties. The two most common are the fairly large, dark-green cucumbers, which are usually waxed to protect them against moisture loss, and the smaller ones, often called Kirbies, which are usually not waxed. The huge ones sealed in plastic film are rarer and costly, and they are variously called European, English, seedless, burpless, hothouse, and greenhouse. Cucumbers are a warm-weather crop, but they are on the market all year with the greatest availability and highest quality in summer. Look for firm fruit with good green color. Light speckles and bumps are to be expected, but white or yellowish areas are an indication of advanced age–they may taste stale or bitter and the seeds may be overgrown, woody, and therefore wasteful. The smaller ones of each type are usually tenderer and better. Cucumbers are very susceptible to chill damage, which shows up as wet-looking areas. Avoid these, along with soft spots (check the ends particularly) and shriveled skin. Store cucumbers, closely wrapped, in a cool place or the produce bin of the refrigerator. Use them within a few days.

Cucumbers need not be routinely relegated to the salad bowl. They also make a refreshing salad of their own when marinated in vinegar. Dill, parsley, and mustard are complementary flavors. An ancient preparation still in daily use is cucumbers combined with yogurt and perhaps mint to form a cooking condiment to go with spicy Arabic or Indian food. And chunks of lightly salted cucumber are standard garnishes with the savory Indonesian skewered tidbits called satay. In classic French cookery, cucumbers are often steamed or sautéed in butter, most often to be served with fish. Of course, cucumbers can be pickled in a variety of ways. The small ones, Kirbies, are sold as pickling cucumbers, but they are delicious for all other uses. Cucumbers turn up in classic English tea sandwiches, and as boats holding the likes of sour cream and caviar. Excess wax can be removed by rinsing with very hot water and wiping dry, but you still might want to peel the waxed ones, at least partially. Peeling removes a lot of the vitamins, but it also cuts down on digestive problems.

WHAT TO LOOK FOR: plump, dark green, smaller ones of that type

WHAT TO AVOID: watery or soft spots, shrivel, yellow or white coloration

HOW TO STORE: wrap closely and keep cool, for a few days

PRIME SEASON: (Indicated by darker shade).

| JAN | FEB | MAR | APR | MAY | JUN | JUL | AUG | SEP | OCT | NOV | DEC |

Dandelion

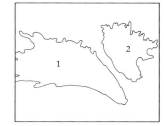

1. Catalonian chicory
2. dandelion

Have you ever noticed how much more reliable folk wisdom is than the pronouncements of scholarly botanists? When the folk call a weed piss-a-bed (or *pissenlit*), the odds are in favor of diuretic tendencies. Even so, dandelion has had a sound reputation for centuries. It has been much enjoyed for its flavor as well as its medicinal properties. The name comes to us not from the charming golden mane suggested by the flowers, but from the spiked leaves, which reminded the French of lion's teeth – *dents de lion.* Gardeners jealous of a perfect carpet of lawn have attempted to stamp it out, while cooks have eagerly gathered it while they might, especially in early spring. Cultivation is under way, so that the greens are increasingly available in the market. They have a rich green flavor and appealing bitterness.

Commercial dandelion is now available year round in limited quantities. For the most delicate flavor, choose very fresh-looking leaves that are bright but not too deep green. They should, of course, be free of wilt, rot, and yellowing. Leaves harvested with the full stems and the root base intact stay far fresher than those that are cut. Use promptly or refrigerate, wrapped, for a day or two if necessary. Much of the dandelion on the market in cold weather is not dandelion at all but Catalonian chicory, a perfectly acceptable substitute that is distinguished by its larger leaves.

After they are carefully washed to remove all traces of grit, dandelion leaves make excellent salad fare, most often in a preparation that includes crisp bacon or some other cured pork. They are also cooked like other greens and are reputed to be successful in any recipe calling for spinach. The flowers are used to make wine and the taproots have been roasted for a coffee substitute like chicory. The leaves can be grown in darkness to produce a pale, delicately flavored result that is similar to Belgian endive.

WHAT TO LOOK FOR:
bright green leaves with stems and root base intact

WHAT TO AVOID:
wilt, rot, and yellowing

HOW TO STORE:
refrigerate wrapped, for a day or two

PRIME SEASON: (Indicated by darker shade).

| JAN | FEB | MAR | APR | MAY | JUN | JUL | AUG | SEP | OCT | NOV | DEC |

Dates

The Western notion of a desert oasis dominated by graceful palm trees is remarkably accurate, for date palms will bear a good sweet crop only when they can sink their roots deep into a reliable water source while thrusting the leaf crown up into the parched air above. The tree grows in many warm places, but it delivers the goods only when humidity is absent. Dates are truly desert manna, rich in life-sustaining energy (and spirit-lifting energy, too, when fermented). The mainstay of the desert has become a popular treat in all climes, usually in its dried state but occasionally fresh as well. Fresh dates taste light but satisfying, without the cloying sweetness that develops when drying concentrates their considerable natural sugars.

Fresh dates are something of an oddity in parts of the world where they are not grown–that is, in most parts of the world. The usual marketing season, if you can find them, is late summer through fall, though they might be available into midwinter as well. Choose dates that are plump and glossy. Do taste to see that you are purchasing soft dates rather than hard, or bread dates, which are about as exciting as the packaged dried fruit mixtures and cheap baked goods for which they are used. Avoid shrivel, mold, stickiness, and any fermented odor. Because of their high sugar content they should keep well when wrapped and refrigerated, at least a week or two if very fresh at purchase.

Dates make a delicious and satisfying (if fattening) snack, as well as a pleasant breakfast or dessert fruit. The fresh ones can be substituted for dried in baking, with lighter results. Fresh dates make a charming change of pace in salads and desserts, where they blend well with other fruits. There is a tradition for stuffing dates–with cream cheese or marzipan or almonds, even wrapping them with bacon and broiling them for hors d'oeuvres–and the fresh ones are an excellent substitute for the overly sweet dried varieties.

WHAT TO LOOK FOR:
plump, glossy, soft

WHAT TO AVOID:
shrivel, mold, stickiness, fermented odor

HOW TO STORE:
wrap and refrigerate, about two weeks

PRIME SEASON: (Indicated by darker shade).

| JAN | FEB | MAR | APR | MAY | JUN | JUL | AUG | SEP | OCT | NOV | DEC |

EGGPLANT

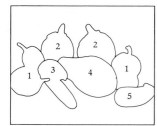

1. Yellow

2. White

3. Chinese – the distinctive, long thin shape and purple color is evident in the market, though not so obvious here.

4. Large purple

5. Italian (small purple)

Though never quite so feared or maligned as its cousin the tomato, eggplant has had a long, slow struggle for acceptance, and remains to this day an important vegetable only in India, where it probably originated, the Mediterranean, to which it migrated, and China, where it may have had an independent start. Eggplant is a strictly American name, a description of the shape of the fruit when it first arrived in the colonies. (At that time it might perhaps have been white.) The French and English name, aubergine, comes from early Hindustani and Arabic words that transliterate to something like *batanjaan*, but the modern Italian and Greek names (*melanzane* and *melitzanes*, respectively) come from a cruel bit of taxonomical slander – a sixteenth-century classification was *mala insana*, 'mad apple,' suggesting the alleged result of ingesting the thing. But that has not stopped Italians and Greeks from being major consumers of this versatile vegetable. Eggplant is meaty and satisfying, distinctive enough to be enjoyed for its own sake, and mild enough to be flavored in myriad ways.

The familiar large purple eggplant is available in nearly uniform supply throughout the year. The small purple ones, usually called Italian, and the long, pale purple Chinese eggplants are also grown all year, but in much smaller quantities. The others, in white, yellow, or grey, are rare and are usually limited to fall. Look for plump, glossy, heavy fruit, whatever its shape or color, and smaller is better. The cap should be reasonably fresh-looking. Avoid dull or shriveled skin and any bruises or rot. Eggplants can be wrapped and refrigerated for a few days if necessary. There is a long-standing controversy about male and female eggplants, an inaccurate approach considering that fruits are the

product of sex and do not have it. However, there is some folk wisdom that is worth attention – eggplants have a dimple at the blossom end that can be very round or oval, and the oval ones seem to have fewer seeds and meatier flesh.

There is another issue: whether or not to salt and drain eggplant before cooking. This practice grew out of two factors – the heavy moisture content that can make a dish watery, and a certain bitterness that is leeched by the salt treatment along with the moisture. Though this bitterness is largely bred out of modern eggplants, it does appear occasionally and most cooks continue the salting ritual. Eggplant can be baked, stuffed, broiled, roasted, sautéed, fried, or stir-fried. The most usual partners are lamb, pork, veal, chicken, cheese, seafood, tomatoes, onions, garlic, peppers, squash, herbs, breading, olive oil, and sesame oil or paste (tahini).

WHAT TO LOOK FOR:
plump, firm, glossy; fresh-looking cap

WHAT TO AVOID:
dull or shriveled skin, bruises or rot

HOW TO STORE:
wrap and refrigerate for a few days if necessary

PRIME SEASON: (Indicated by darker shade).

| JAN | FEB | MAR | APR | MAY | JUN | JUL | AUG | SEP | OCT | NOV | DEC |

FENNEL

Another Mediterranean vegetable has begun to find a wide audience. Fennel was enjoyed by the ancient Romans and prized throughout medieval Europe as vegetable and as medicine. Folk wisdom scores again, for it was much used as a treatment for eye ailments, and modern chemistry has determined it rich in vitamin A. Whatever its healthful properties, fennel deserves popularity for its refreshing crispness and subtle aniselike flavor. The table variety is called sweet, Florentine, or Roman fennel, or by its Italian name *finocchio* (a word to use with care because it is also a slang term for gay men). This variety has a thick bulbous stem base (the most prized part for eating), celerylike stalks that can be used like celery, and feathery leaves that can be used as a green herb. In the south of France another variety of fennel is grown for its leaves, for its seeds, and for its stalks, which are most often used fresh or dried as fuel for grilling fish, thus imparting a delicious flavor.

Fennel, sometimes called anise in the market (even though it has no relationship to that plant), is available in all seasons but summer. November and December are peak months. Choose fat white bulbs of three or four inches in diameter with fresh green leaves and stems, preferably not trimmed to less than eight inches long. Trimming robs one of edible parts and is also a good indication that wilt is being disguised. Avoid rot, bruises, and splits. Good fresh fennel can be stored in the refrigerator, wrapped, for a few days.

The leaves of Florentine fennel can be used in salads or for flavoring soups and stews. Here is a fine substitute for French fennel, an essential ingredient in a good *bouillabaisse.* The stems may be munched like celery or sliced into salads. The crisp white bulb has been blanched by banking it with soil, so the flavor is especially delicate. Slice lengthwise for salad or hors d'oeuvres. It can be cooked in the traditional French manner *à la grecque*– poached in a spicy liquid that is then reduced (boiled down) and poured over the vegetable and left to chill and marinate. Fennel is sometimes sliced in half or quarters lengthwise and braised. *Gratinée* with cheese is an excellent preparation. Cooked fennel is most often used to flavor or accompany fish dishes, but it is equally delicious with other flesh, especially broiled or roast veal and chicken.

WHAT TO LOOK FOR:
fat, round white bulbs with fresh green leaves and stems

WHAT TO AVOID:
rot, bruises, splits, wilt, excessive trimming

HOW TO STORE:
refrigerate wrapped for a few days

PRIME SEASON: (Indicated by darker shade).

| JAN | FEB | MAR | APR | MAY | JUN | JUL | AUG | SEP | OCT | NOV | DEC |

FIDDLEHEADS

This fiddlehead came from a group that was fuzzier than usual, and the little brownish leaflet sprouts should be picked off. You will save yourself time if you can find smoother ones.

Fiddleheads are among the most charming of exotic vegetables. Each spring the ostrich fern of New England and Canada (so named because its shape resembles the bird's head and neck) sends up delicate green shoots that are indeed reminiscent of the tuning-peg end of a violin. Biology students may remember (or not) circinate vernation, a term describing the way fern leaves begin life tightly rolled into a coil, unrolling and spreading their lacy lobes as they mature. The trick here is to catch the shoots while they are very young and between only two and eight inches long, with the tip still rolled up like a crosier. At this stage they have a delicate flavor that is a bit asparaguslike with woodsy tones that remind some of mushrooms. The Indians of the Northeast discovered this delicacy long ago.

The season is short, usually May only, the price is high, the distribution is limited, but fiddleheads are worth seeking in specialty stores. Choose only fresh young shoots of the smallest size. Avoid wilt, rot, and overgrown specimens that have begun to unfurl. Use as soon as possible. Frozen and canned fiddleheads are not very interesting.

Fiddleheads must always be washed carefully and gently. They may be eaten raw, as salad with a lemony vinaigrette. A quick sauté in butter with a grinding of pepper and a splash of lemon juice is a fine preparation. Steaming is a good idea, but only for a few minutes. Serve as you would asparagus, with lemon butter, *sauce hollandaise*, or *sauce maltaise*.

WHAT TO LOOK FOR:
fresh young shoots, tightly furled ends

WHAT TO AVOID:
wilt, rot, any over eight inches long

HOW TO STORE:
refrigerate wrapped, unwashed, for a day

PRIME SEASON: (Indicated by darker shade).

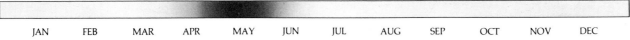

| JAN | FEB | MAR | APR | MAY | JUN | JUL | AUG | SEP | OCT | NOV | DEC |

FIGS

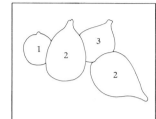

1. Smyrna (Calimyrna)
2. Black mission
3. Green mission

Figs have never been taken lightly in their long association with mankind. Such a valuable food source was sacred to the ancients along the Mediterranean, and the fig's voluptuous fecundity has been a source of dirty jokes ever since. Spanish explorers brought figs to the New World in the early sixteenth century, as did Spanish missionaries. They found a providential niche (the figs and the missionaries) in California, which now produces one of the world's major crops. The New World version of the black (or purple) fig is Mission, named for where it first grew. There is also a green Mission. Another important fig is the Smyrna, named for the ancient Turkish city, now called Izmir, where it has thrived for centuries. This round, green-gold beauty is called Calimyrna in the U.S. Kadota is the other major commercial variety, green-gold like Smyrna but slightly more pear shaped, like the Missions. When fully ripened, they are all equally juicy, sweet, and irresistible.

There are two seasons for figs, an early one in June and a late summer season that continues until the first frost. The best supply is in August, September, October, and early November. Choose plump fruit that is fresh-smelling and soft to the touch. Hard, underripe fruit will not improve much with time. Figs are highly perishable, so avoid blemishes, shrivel, rot, mushiness, and any moisture beyond a

dewdrop or two at the base end, the ostiolum. Be especially careful to avoid a fermented odor, sour or musty, which occurs in overripe figs that have begun to weep puddles of juice. Use figs promptly or store them, covered, in the refrigerator for a day or two. Wash gently just before serving.

Figs are used fresh, dry, and canned in a wide variety of preparations, but the fresh ones are at their luscious best when served simply. They are especially good as hors d'oeuvres with thinly sliced prosciutto or other fine raw ham. Cream is a classic accompaniment for breakfast or dessert. Eaten out of hand, the entire fig is certainly edible, but some people prefer to cut or tear it open and eat only the juicy, pleasantly grainy center.

WHAT TO LOOK FOR:
fresh-smelling, fairly soft

WHAT TO AVOID:
blemishes, shrivel, rot, mushiness, fermented odor, puddles of juice

HOW TO STORE:
keep at room temperature or refrigerate, a day or two

PRIME SEASON: (Indicated by darker shade).

| JAN | FEB | MAR | APR | MAY | JUN | JUL | AUG | SEP | OCT | NOV | DEC |

GARLIC

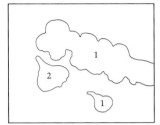

1. White garlic
2. Elephant garlic

It is powerful medicine indeed that cures anything from plague to the common cold, that repels vampires and delights dogs. Such is the reputation (that dogs love it is verifiable) of the most controversial member of the onion family. Garlic has been creating factions since the ancients, for it seems that a lukewarm response to its piquant flavor is as rare as flying pigs. Even those who loathe the taste of garlic have to admit to its medicinal properties; garlic is rich in sulfurous compounds with proven antibacterial powers, still in pharmaceutical use today. Those who dote on it make no apology for this delightful herb with a rich sunny flavor ranging from warm and sweet to hot and pungent, depending upon how it is used.

Fresh garlic is a dried bulb with papery skin that should feel firm, heavy, and full. Avoid bulbs (also called heads) with soft cloves or any rot or mildew. Garlic ranges in color from very white to yellowish to purplish, and there is a size range as well. They are all interchangeable, but many people feel that the smaller ones have finer flavor. Elephant garlic is a fairly new type that has yet to prove its virtues, beyond the fact that its huge cloves are easier to peel than tiny ones. Many cooks believe that it takes the same number of cloves of any garlic to properly flavor a dish, so the seeming benefit of size alone is not a time saver. Garlic in good

condition should keep very well, a month or so, if stored in a cool, dry, well-ventilated place.

Garlic figures prominently in the cooking of much of the world, especially in warm climates. Only in colder parts – most notably North America, Britain, and Scandinavia – is its use fairly limited in traditional cookery. Garlic lovers relish it in everything except ice cream. To be precise, garlic is compatible with all foods but sweets.

Garlic's flavor varies with the way it is physically manipulated and the way it is cooked. Whole, unpeeled garlic is mildest, while garlic mashed to a paste is most pungent; simmering gently in liquid will produce the most delicate result, while sautéing in hot fat produces the most assertive flavor (garlic should never be browned beyond a light golden color, for its flavor deteriorates rapidly when overheated). The easiest way to peel garlic cloves is to crack the skin with the side of a heavy knife. This loosens the skin and makes it easy to remove. Most good cooks feel that garlic should be prepared with a sharp knife or a mortar and pestle – garlic presses and food processors tend to overwork it and produce an off flavor that is usually described as bitter.

WHAT TO LOOK FOR:
firm, heavy, full bulbs

WHAT TO AVOID:
softness, rot, mildew, sprouting

HOW TO STORE:
cool, dry, well-ventilated place for a month or more

PRIME SEASON: (Indicated by darker shade).

| JAN | FEB | MAR | APR | MAY | JUN | JUL | AUG | SEP | OCT | NOV | DEC |

GINGER

The Western world would not be quite the same without gingerbread, nor would holiday baking in general have the same zest without the warmth that tropical ginger brings to it. The holiday relationship began in the Middle Ages, but the Western love affair with ginger reaches back to the ancient Greeks. The Romans were also avid consumers of ginger, and though the fall of the Empire put a considerable damper upon trade with Asia, ginger seems to have gotten through, even before Marco Polo repopularized the wonders of the Orient. Though firmly established in Europe and the U.S. for centuries, ginger has often been thought of merely as a dry powder to be combined with others for baking, the flavoring in ginger ale, or an ingredient in exotic relishes and chutneys. The possibilities for fresh ginger root in daily kitchen use are now being explored, after the example of Asian cooks. Fresh ginger is wonderfully aromatic–sweet and hot at once. For those who dote on it, a little is good and a lot is better.

Ginger root is available year round, always growing in some tropical climate. It may be fresh with very thin, fragile skin, or drier and sturdier though still moist inside. The two states are nearly identical in flavor, though young ginger is more pungent.

Choose fat 'hands' with the least number of knobs and small branchings–little parts are difficult or impossible to peel and may end up discarded. Avoid cracks, shriveled parts, and any rot or mold. Test for a fresh smell. Whole ginger root keeps well when loosely wrapped and refrigerated, for a good week or two. It may also be frozen and sliced off as needed. A convenient and effective storage method is to peel the root and refrigerate it, sealed, in dry sherry to cover. It should keep for up to a month, and the sherry can be reused for storage or used in cooking.

In addition to powdered ginger (which has a mild flavor), there are also crystallized and preserved ginger, which are tasty sweets and work well in cooking, and pickled ginger, which is a standard garnish with Japanese sushi and sashimi. Fresh ginger root is best known to Westerners as an essential ingredient in Asian cookery. Indeed it is indispensible for myriad dishes of Chinese, Japanese, Korean, Southeast Asian, Polynesian, and Indian origin. And ginger is relatively popular in the Middle East and the Mediterranean. Even more versatile than its frequent partner garlic, ginger flavors sweets as well as savories. Modern cooks in the West are teaming ginger root with just about anything, producing traditional Asian foods and new dishes as varied as filet of beef in ginger sauce and ginger soufflé.

WHAT TO LOOK FOR: smooth skin with a slight sheen, large 'hands'

WHAT TO AVOID: shrivel, rot or mold, numerous knobs, musty odor

HOW TO STORE: refrigerate loosely wrapped, freeze, or peel and cover with dry sherry

PRIME SEASON: (Indicated by darker shade).

| JAN | FEB | MAR | APR | MAY | JUN | JUL | AUG | SEP | OCT | NOV | DEC |

GRAPEFRUIT AND UGLI FRUIT

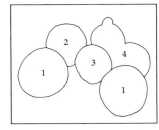

1. Red (Ruby Seedless)
2. Pink
3. White
4. Ugli fruit

Certainly an odd name for such a big yellow thing. Grapefruit is a nineteenth-century name that seems to come from the way the fruit grows in bunches. But one misguided horticulturist actually suggested that they taste like grapes. Wrong. It is a full-out, sunny, citrusy flavor: sweet, tart, and bitter at the same time. The grapefruit's ancestor is an ancient Asian fruit called pummelo or pomelo. No one knows for sure whether the grapefruit happened by accident after pummelos reached the West Indies, or whether it is an engineered hybrid. Legend has it that a seventeenth-century British sailor named Captain Shaddock brought the first pummelos to the West, and whether or not he actually did, they were called shaddocks long before they came to be grapefruit.

Ugli fruit, on the other hand, is very accurately named. It may well be the ugliest of delectable fruits. Its origins, too, are shrouded in mystery, but we can be reasonably safe in suggesting that it is the result of a cross between a grapefruit or pummelo and an orange or tangerine. Whatever the parents, baby got all the best flavor genes and the worst possible appearance.

Grapefruit and uglis are at their best in midwinter, though grapefruit is available all year. Uglis will be fairly green skinned, touched with yellow, but avoid shrivel and browning. Grapefruit should be plump and heavy, round, and either bright yellow or yellow shot with pink. Avoid lightweight or squashed fruit, and shrivel or browning. Pink and red grapefruit are pretty and sweet, but most authorities assert that the yellow ones are better. Most modern grapefruit is fairly seedless. Grapefruit and uglis should be refrigerated and eaten within a few days.

Ugli fruit is delicious eaten out of hand, though it can be used for fruit salad or other desserts. Grapefruit is more versatile. It is satisfying for breakfast, as fruit or juice, and it combines with other fruits or with poultry or pork for other meals. Grapefruit has been broiled and used for sherbet and ices, and even a rich mousse flavored with Pernod.

WHAT TO LOOK FOR: plump, heavy, round grapefruit, fresh-looking ugli with good color

WHAT TO AVOID: shrivel, browning, softness, lightweight grapefruit

HOW TO STORE: refrigerate for a few days

PRIME SEASON: (Indicated by darker shade).

	JAN	FEB	MAR	APR	MAY	JUN	JUL	AUG	SEP	OCT	NOV	DEC
Grapefruit												
Ugli fruit												

GRAPES

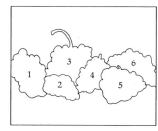

1. Muscat – seeds, very sweet with an interesting flowery perfume.

2. Concord – seeds, rich 'foxy' flavor.

3. Emperor – seeds, pleasant flavor, one of the most popular.

4. Thompson Seedless – the most popular table grape, sweet.

5. Ribier – seeds, pleasant flavor.

6. Red Seedless – one of the new varieties that are gaining popularity, pleasant but unremarkable.

There are other varieties that replace the major ones off-season, and a number of fine regional grapes that do not travel.

If Dionysus had not introduced mankind to the pleasures and sorrows of the vine, we might still be struggling along anyway–the gods know that anything with some sugar or starch can be, and often is, fermented. But whole cultures are shaped by their eating and drinking habits, and the West is decidedly viniferous. Old World grapes in their many forms are the most important ones, supplying all the fine wine grapes and most table grapes as well. There is also the native American grape, so plentiful in the past that Leif Eriksson was prompted to name his discovery Vinland. American grapes are of the sort that the English colonists called fox grapes, perhaps because of their slightly musky flavor, or because the name makes them sound charmingly wild. Muscadines and Scuppernongs in the South are representative of the native type, as are cultivated and 'improved' Concords and Catawbas in the North. Some do make it to market as table grapes, though they do not keep or travel well. Distribution is usually limited to neighboring areas, and the season is short–late summer and early fall only.

The major table grapes are available most of the year because of modern growing and storage techniques and imports. They will generally be best in late summer and early fall, for this is a one-harvest crop that takes the entire growing season to develop its flavor and sweetness. Fruit picked prematurely will not ripen. Grapes that have not been mishandled have a slight frosty film on the skin called bloom. There should be the barest minimum of bruised, blighted grapes in each bunch, and good bunches really do look the way tradition dictates–full at the top, tapering into a soft conical shape. Do not wash until ready to use, for moisture hastens decay.

Fine table grapes need no enhancement, but they do marry well with rich dairy products for dessert. The American version blends green grapes with sour cream and brown sugar. In Italy, grapes might be served with the rich cream cheese *mascarpone*, perhaps flavored with rum and a little cinnamon or other sweet spice. Seedless grapes are very good in fruit tarts and salads. Other grapes should be halved and seeded. For such preparations it is tedious, but polite, to peel the grapes, starting at the stem end. Remembering Mae West ('Beulah, peel me a grape') may help to relieve the tedium. Seedless green grapes are occasionally combined with a rich but delicate sauce for seafood or poultry. In classic French cooking, the grapes give the preparation its name, *Veronique*. Dedicated home cooks still create their own jams, jellies, preserves, and even wines from grapes they grow or purchase.

WHAT TO LOOK FOR:
plump, firm fruit; deep color for black and red, yellowish for green; fresh, greenish, pliable stems, 'bloom'

WHAT TO AVOID:
bruises, blight; dry, brittle, brown stems; moisture; loose grapes; softness and decay; whitish areas under the skin

HOW TO STORE:
wrap loosely and refrigerate, unwashed, for a few days if necessary

PRIME SEASON: (Indicated by darker shade).

| JAN | FEB | MAR | APR | MAY | JUN | JUL | AUG | SEP | OCT | NOV | DEC |

GREENS

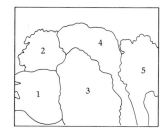

1. Kohlrabi
2. Kale
3. Swiss chard
4. Collards
5. Rape

Potherb is a charmingly old-fashioned word and it covers a lot of territory indeed. It denotes all those leafy things, plus their stems, that we throw into a pot and cook up into what is known in the American South as 'a mess of greens.' As such they are a cornerstone not just of 'soul food' but of most diets around the world. Many greens are members of the cabbage family–kale, collards, rape (also known as rabe, *broccoli* or *broccoletti di rape,* and *rapini*), kohlrabi, turnips, mustard greens, and broccoli or cauliflower leaves. Others are beets–beet greens and Swiss chard. Dandelion and other weeds fit this category, along with familiar spinach, lettuces, and chicories. And there are all manner of other things, wild and domesticated, that are or have been tossed into the pot for their own sake or to flavor other foods. Despite their botanical differences, greens are similar in flavor. They have an intense, unmistakable vegetable taste with varying degrees of bitterness that delights those who dote on them and repels those who do not.

Greens are available year round, many at their best in cold weather. Choose bright green, crisp, fresh-looking leaves. Avoid wilt, yellowing, and rot. Insect damage is fairly common and easily trimmed away if slight, but avoid greens that are riddled with tiny holes. Fresh is always best, but greens in good condition can be refrigerated, wrapped, for a few days if necessary.

Greens are often served just for themselves with possible additions of butter, olive oil, garlic, onions, salt pork, bacon or smoked ham, or an offering of vinegar (splashed on to taste), a brisk flavor addition that complements the natural bitterness. They can marry well with any flesh, especially spicy preparations, roast pork, and fried fish. Rice, beans, corn, and potatoes are usual starch accompaniments. Greens also flavor soups and stews made with meats and poultry and all kinds of vegetables. Greens preparations may also include cheeses, nuts, eggs, and cream or white sauces. Greens that have been bred for their roots–beets, turnips, and kohlrabi–are sometimes combined with the root, yielding a variety of colors, flavors, and textures in the same dish.

WHAT TO LOOK FOR:
crispness, bright green color

WHAT TO AVOID:
wilt, rot, yellowing, and severe insect damage

HOW TO STORE:
refrigerate wrapped, unwashed, for a few days

PRIME SEASON: (Indicated by darker shade). This may vary slightly due to local crops.

| JAN | FEB | MAR | APR | MAY | JUN | JUL | AUG | SEP | OCT | NOV | DEC |

HERBS

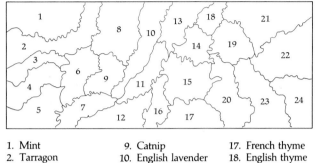

1. Mint
2. Tarragon
3. Greek oregano
4. Lemon thyme
5. Lemon balm
6. Bay
7. Summer savory
8. Coriander
9. Catnip
10. English lavender
11. Rosemary
12. Chervil
13. Salad burnet
14. Woodruff
15. Golden sage
16. Tri-color sage
17. French thyme
18. English thyme
19. Dwarf sage
20. Borage
21. Rose geranium
22. Marjoram
23. Lime geranium
24. Chamomile

Little green plants have been a mainstay of human life since its beginnings. A staggering variety of herbs have been grown or gathered for their flavor, preservative qualities, nutritional value, real or imagined medicinal properties, and even magical powers. Every culture has its favorites, based originally on what grew where, now so deeply ingrained that the savors and scents of fresh and dried herbs almost define a people. There are exceptions, but the general rule is that herbs grow in temperate climates, while spices are tropical; herbs are likely to have useful leaves, stems, or flowers, while spices are more often grown for their seeds, berries, roots, bark, or other parts, often dried before use.

Spring and summer would seem to be the best times to find fresh herbs, and in some areas this will prove true. This is also the best time to find potted herbs to grow at home. But in large urban areas, parsley, mint, and dill, even coriander and basil, should be available daily, while other herbs are also being shipped from warmer growing areas or hot-housed locally for sale throughout the year. One need only find a market that carries them. Look for the freshest possible leaves. Avoid wilt, dryness or browning around the edges, and bruising, yellowing, or rust. Always check herbs for a fresh scent and bruise a leaf to see that the characteristic odor is strong and clean. There are varieties of some herbs on the market occasionally that have little or no flavor, so you must know how that herb is supposed to smell and taste. Fresh herbs should keep well for a day or two refrigerated. For longer keeping, up to a week, place the stems in a glass of water and cover the leaves with a plastic bag secured to the glass with a rubber band or string. Refrigerate and change the water daily.

ANGELICA
This is most visible as candied stems dyed a vivid green and used by bakers, but the leaves and roots lend an interesting pungency to fish preparations and stewed fruits.

BAY LEAVES
Either fresh or dried, they are indispensable in the herb bouquet that flavors so many Western soups, stews, stocks, sauces, and casseroles. Use sparingly.

BORAGE
This makes a refreshing, cucumberlike salad addition when the leaves are very young. Older leaves can be used as a potherb, while the pretty blue flowers can be eaten raw or crystallized for confections.

CATNIP
It is also called catmint and is a strongly flavored mint that cats (and other felines) really do like, as do humans, who usually brew the leaves or flowers into a tea.

CHAMOMILE
This is a daisylike plant whose flowers are most often used as a sedative tea, while the leaves have some of the same calming effect.

CHERVIL
It is an essential *fine herbe* with a delicate parsley-like flavor that has never quite caught on in the U.S.

CORIANDER
Also called Chinese parsley or cilantro, this is an essential ingredient in Asian and Latin American cookery with a pungent odor and flavor that startle the uninitiated. Coriander got its name from the ancient Greek *koris,* 'bedbug,' a reference to the smell.

PRIME SEASON: (Indicated by darker shade).

| JAN | FEB | MAR | APR | MAY | JUN | JUL | AUG | SEP | OCT | NOV | DEC |

DILL
With its familiar, feathery leaves, dill is most prized in Scandinavian and Eastern European cookery, where its sprightly flavor perks up vegetables and seafood. An oil pressed from the seeds flavors pickles.

SCENTED GERANIUM LEAVES
They come in many flavors, including rose, lemon, lime, and peppermint and are used to scent or flavor sweets, finger bowls, and potpourri.

LAVENDER
It is most often thought of as a medicinal or cosmetic, but the leaves can be used in marinades for game, and the flowers may be jellied or crystallized.

LEMON BALM
This is tasty, both lemony and minty, for savories and sweets and, especially, tea.

LEMON GRASS
This is a lemony grass that is essential in Southeast Asian cookery.

LEMON VERBENA
It does indeed have a lemony odor and taste that can be used wherever lemon would be welcome. Lemon verbena tea is reputed to be good for calming the nerves, settling the stomach, and treating colds.

LOVAGE
This is a large celerylike plant with a pungent celerylike flavor that is welcome in hearty soups, stews, casseroles, and salads.

MARJORAM
This has a warm, Mediterranean flavor that is best suited to poultry and veal, salads, eggs, and vegetables. It is botanically almost identical to oregano and tastes it.

MINT
In the market you may find one of several varieties of this herb that vary in their pungency. Taste will govern the quantity you add to sweets, fruits, relishes, lamb or poultry preparations, vegetables (especially tomato sauces and squash), cucumbers and yogurt, or tea.

OREGANO
This is a pungent member of the marjoram group, at its best with seafood, poultry, and tomatoes.

ROSEMARY
Used with care to flavor poultry, veal, seafood, and marinades for game, rosemary has also been used to flavor sweets and jellies.

RUE
This is used occasionally, in small quantities, to flavor salads and cheeses, and often to turn *grappa*, the famous Italian marc, into *grappa al ruta.*

SAGE
It is most often used to flavor fatty meats–pork, goose, and duck–and cheese. It is a common 'poultry seasoning' for stuffings. Sage comes in several varieties.

SALAD BURNET
This has a refreshing bitterness and coolness that makes it pleasant with salads, cheese, vegetables, soups, and beverages.

SAVORY
Whether the summer or the winter variety, this has an interesting herbal flavor more like thyme than anything else and is welcome in sausages, beans, stuffings, and meat pies.

TARRAGON
This is the queen of French herbs, its green herbalness and slight anise flavor lending a subtle note to sauces, butters, eggs, salads, and seafood. It is often used to flavor vinegars.

THYME
In its various forms, this is probably the most ubiquitous herb after parsley, at home in a wide variety of savory preparations.

WOODRUFF
It is the characteristic herbal flavor in May wine. It can also be used for sherbets and other sweets.

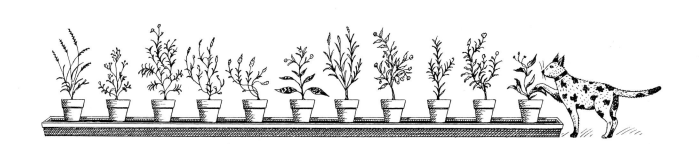

WHAT TO LOOK FOR: freshest possible leaves; clean, characteristic scent

WHAT TO AVOID: wilt; dryness or browning around the edges; bruising, yellowing, rust; off odor

HOW TO STORE: wrap in damp paper towel for a day or two, or place stems in water in a glass, cover, and refrigerate for up to a week

HORSERADISH AND WASABI

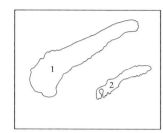

1. Horseradish
2. Wasabi

The pretty little sprouts on the wasabi are not ideal, but no major problem so long as the root is firm and crisp.

Horses probably do like it. Humans certainly do, especially if they are German, French, English, or American. Horseradish is a very popular condiment, also one of the five bitter herbs of the Passover seder. It has a delightful kick to it, and is at its very best when fresh, rather than bottled. A dried version exists too, and it is worse than the bottled stuff. The Japanese have a root with a similar hot flavor, called wasabi or Japanese horseradish. This one is often processed into a green powder to be mixed to a paste with a little water, lending sinus-clearing heat to sushi and sashimi. Occasionally it can be found fresh in Asian markets.

Horseradish is available year round in small supply, especially plentiful in the spring, less so in late summer. Wasabi too is available most any time, if you can find it at all. The roots are always gnarled and warty, but they should be plump, crisp, and fresh-looking, not dry and shriveled. Leaf sprouts are not ideal, but they are acceptable so long as the root is still firm. Refrigerate and use within a few days.

Horseradish makes a bracing nibble at hors d'oeuvre time, but it is most often finely grated to be combined with cream, sour cream, mayonnaise, or tomato-based condiments for a sauce. The creamy sauces are served with roast beef, boiled beef, and sometimes tongue, lamb, or even fish. They also enliven salads, especially potato. The red stuff is used as a cocktail sauce for cold shrimp,

oysters, or crab. If you find horseradish marketed with fresh leaves attached you can use the leaves, sparingly, in salads. Horseradish makes an interesting pickle. Wasabi is also at its best when freshly grated to be served with Japanese dishes, or pickled. Both horseradish and wasabi lose flavor rapidly and should be grated as close to serving time as possible. Heat destroys them, so they are always used raw.

WHAT TO LOOK FOR:
plump, firm, crisp roots

WHAT TO AVOID:
wilt, shrivel, soft spots, sprouting

HOW TO STORE:
refrigerate and use within a few days

PRIME SEASON: (Indicated by darker shade).

| JAN | FEB | MAR | APR | MAY | JUN | JUL | AUG | SEP | OCT | NOV | DEC |

KIWI FRUIT

No one knows for sure how it came to be called Chinese gooseberry–it probably is not Chinese–but the kiwi part can be explained. Clever New Zealander promoters realized they were up against the American cold war mentality, so they deleted the offending reference to what was then a feared country, China, and substituted a charming, safe reference to the national bird of New Zealand. Thus was born the kiwi fruit, and it flourished. Few edibles have had such a meteoric rise to international celebrity, and it can only be hoped that this one will survive its fifteen minutes of stardom to become a kitchen staple. And many people who really like it pray for the day when it loses its trendy nouvelle cuisine stigma altogether so that they can enjoy it without embarrassment. Good ripe kiwi is charming to look at, and the pulp has an excellent sweet/tart balance. The distinctive fruitiness is difficult to describe, but at least a few people have found that it reminds them of fruit-flavored chewing gum refined to the ultimate degree.

Kiwi fruit is now available all year, no longer the exclusive product of the growing season in New Zealand. Choose fruit that is slightly soft to the touch and store it at room temperature for a few days, loosely wrapped, until it becomes fairly soft and pleasantly scented but by no means mushy. This approach avoids the inevitable rough treatment of ripe fruit in the market. Avoid kiwis that are rock-hard or the ones that are very soft, shriveled, or bruised. Once nicely ripened at home, kiwis can be refrigerated for a few days if necessary.

Once the strange furry skin is pared away, kiwis are generally thinly sliced crosswise and used to garnish a variety of dishes–fruit salads, tarts, cakes, pies, and savory concoctions, especially salads made with breast of duck or chicken. They are also attractive, and tasty, when cut into wedges and wrapped with thinly sliced prosciutto or other fine raw-cured ham for hors d'oeuvres. Because of their beauty, and their relative costliness, kiwis are usually featured boldly and visibly on the surface of whatever it is they garnish. However, the price is coming down with increased supply, so they should soon be more frequently seen cut up or puréed in ices and ice creams, jams, jellies, and preserves, mousses, bombes, and soufflés.

WHAT TO LOOK FOR:
plump, slightly soft fruit

WHAT TO AVOID:
rock-hard or mushy fruit, shrivel, bruising

HOW TO STORE:
ripen at room temperature, then refrigerate for a few days if necessary

PRIME SEASON: (Indicated by darker shade).

| JAN | FEB | MAR | APR | MAY | JUN | JUL | AUG | SEP | OCT | NOV | DEC |

KUMQUATS

It is a beautiful little thing; rare and inconsequential. Botanists don't even agree on what it is. It might be a citrus fruit; the name means 'golden orange' in its native Canton. Whatever it is, growers do produce a small crop, and the kumquat remains one of the most charming harbingers of winter. The thick skin is surprisingly sweet, though there is a hint of the bitterness associated with citrus peel. The pulp is intensely sour, so that the kumquat, eaten whole, treats the palate to a wonderful balance of flavors. The taste is reminiscent of orange with a little tangerine thrown in, but the zestiness is distinctly individual.

Kumquats are most often preserved or pickled, but some fresh ones do come to market beginning in mid-fall. They should still be available for traditional Chinese New Year gifts in midwinter, a most pleasant way to wish good luck for the year. Kumquats are usually sold in baskets, still attached to their stems and evergreen leaves. Choose fresh-looking greens and avoid any mold–it spreads quickly in all fruits of this type. Kumquats should be firm and blemish-free with bright skin.

The majority of fresh kumquats are probably used for decoration–little branches of fruit with leaves are visually appealing. For eating raw, kumquats should be uniformly bright orange, while paler greenish ones are successful when cooked for preserves, garnishes, and sauces. One always thinks of duck as a good partner, but the sprightly flavor of kumquats will also give a lift to other fatty meats–pork and goose. Kumquats combine well with other fruits for salads and sweet preparations, and they can be pleasant with chicken, turkey, and game birds.

WHAT TO LOOK FOR: bright, shiny fruit, fresh green leaves if present

WHAT TO AVOID: mold, rot, dry leaves

HOW TO STORE: refrigerate loosely wrapped, for a few days

PRIME SEASON: (Indicated by darker shade).

| JAN | FEB | MAR | APR | MAY | JUN | JUL | AUG | SEP | OCT | NOV | DEC |

LEEKS

Food fashions know neither logic nor justice; if they did, leeks would be among the most honored and precious of vegetables. Indeed, they have been much admired through the ages, credited with varied medicinal powers from providing the strength needed to build pyramids to an antidote for graying hair to a cough remedy. And the leek figures as an emblem of national pride: the Welsh wore it in their caps when King Cadwallader led them to victory over the Saxons in 640, the Irish claim it as a miracle of St. Patrick's, and the Scots dote on cock-a-leekie soup. But the French, who have managed to use it quite wisely in the kitchen, prefer to think of it as 'poor man's asparagus.' Everybody else uses it without much thought, except for Americans, who hardly use it at all. Leeks' fancy price tag in the U.S. indicates lack of interest and small supply, an unfortunate situation for one of the most interesting, sweet, and subtle members of the onion family.

In markets that carry them, leeks should be available all year, at their most plentiful fall through spring. Smaller leeks are usually best, and the ones with the longest white bases are the most prized for usual kitchen purposes. Leeks that have not been trimmed excessively will stay fresh longer and are probably fresher to begin with. Avoid wilt, rot, browning, and yellowing of the leaves. Always examine the center of the leaves to determine that there is no solid, tough flower stalk, an indication of advanced growth. These old leeks will be woody in the center and often betray their age by starting to bulb at the base, while young leeks are essentially straight.

Leeks are banked with soil for blanching, so they are always sandy. They must be washed thoroughly, so it is best to trim off the dark greens (reserving them to flavor soups and stews), then cut the shaft in half or quarters almost down to the root, leaving it attached to hold things together. Once cleaned, leeks are trimmed and prepared in a variety of ways. The white part is traditional in stocks, soups (especially potato – *vichysoisse* when served cold), and various stews and casseroles. Leeks can be braised and served as a hot vegetable or salad or used to flavor steamed mussels or *bouillabaisse.* They complement most any flesh or other vegetable. Thin strips of leek in *julienne* are used to garnish various dishes, including fish steamed in the Chinese style.

PRIME SEASON: (Indicated by darker shade).

| JAN | FEB | MAR | APR | MAY | JUN | JUL | AUG | SEP | OCT | NOV | DEC |

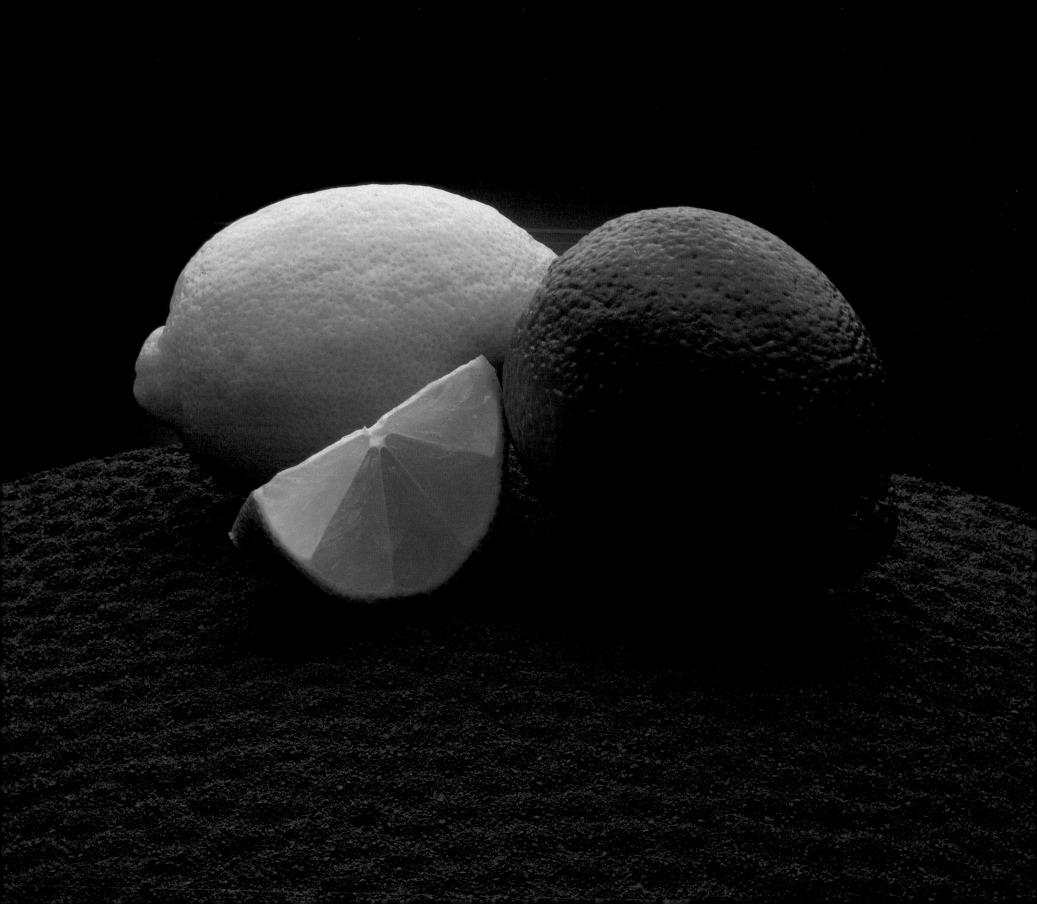

LEMONS AND LIMES

1. Lemon
2. Lime

It is difficult to imagine life without lemons, though Western Europe struggled along for centuries relying upon verjuice (juice of sour grapes) to add the necessary touch of tartness to so many foods. The Greeks and Romans imported lemons from the Middle East, but like so many niceties of civilization, lemon trade was interrupted by the fall of the Roman Empire. Without the lemon, Europe endured Dark Ages indeed, lacking the sunny flavor of this important citrus. The Arab world naturalized lemons and limes and eventually spread them throughout the Mediterranean, whence they gained popularity in Europe and found a happy home in the tropics of the New World. Lemon is a universal favorite, lending its sprightly flavor to a wide variety of cuisines. Limes are hardly an oddity anywhere, yet they have never quite gained the same mass appeal. Beyond limeade (arguably the most sublime of soft drinks) and cocktails, limes are put to best use in the tropics where they grow.

Unlike other citrus fruits with definite seasonal crops, lemons and limes are likely to have blossoms, immature fruit, and ripe fruit on the tree most all the time. So lemons and limes are in good supply throughout the year, most plentiful in summer. The usual lime on the market is a green one called Persian, while the small yellow ones usually called key limes are generally available only where they grow, in Florida and the Caribbean. Choose plump, firm fruit that feels heavy for its size–this is a good indication that it will be juicy. Avoid fruit that is soft, shriveled, bruised, or lightweight. Be very careful to avoid mold, for it spreads rapidly. Lemons and limes in good condition should keep well, wrapped and refrigerated, for a week or so.

Both the juice and the zest (the thin outer colored layer of the peel) are used for flavoring. To juice a lemon or lime most efficiently, have it at room temperature and roll it firmly over a flat surface to break down the pulp. The fingers of one hand bunched together make a dandy reamer for juicing a half-fruit. Remove zest with a grater or with a swivel-blade vegetable peeler, being careful to avoid the bitter white pith beneath. Lemons turn up as a main or complementary flavor in all manner of foods, equally at home with meats, poultry, seafood, dairy products, fruits, vegetables, beverages, and baked goods. They are delicious in both sweet and savory preparations. Limes are just as versatile, but most often used in dishes that reflect ethnicity–Central American, Caribbean, Indian, Southeast Asian, or Polynesian.

WHAT TO LOOK FOR:
firm, plump, heavy fruit, with thin skin

WHAT TO AVOID:
softness, shrivel, bruises, mold

HOW TO STORE:
refrigerate for a week or so

PRIME SEASON: (Indicated by darker shade).

| JAN | FEB | MAR | APR | MAY | JUN | JUL | AUG | SEP | OCT | NOV | DEC |

HEAD LETTUCE

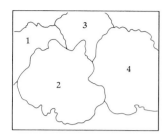

1. Hydroponic
2. Bibb
3. Iceberg
4. Boston (butter)

Bibb lettuce is likely to be well trimmed in the market, and quite tiny compared to Boston. A full head like this one is less usual, and a good value.

A head of lettuce means several things these days. Assuming the cooperation of Mother Nature, at least three varieties of head lettuce should be available any day of the year. 'Crisphead' and 'butterhead' are the two categories. Iceberg is the usual name for the crisp ones. Because of its superior shipping qualities, iceberg is by far the best seller in the U.S., also one of the least tasty and nutritious plants in modern agriculture. But the butterheads are growing in availability and popularity. Although extremely fragile and perishable, they are also tasty. Boston is the American name for the larger one, a popular and usual type of lettuce in Europe too. Lettuce of this sort is also grown hydroponically, supported only by a strip of plastic foam, to produce a pretty, soft green, living lettuce that, unfortunately, has little flavor. The star of the lot is the little gem called Bibb, after the Kentuckian who developed it. It is buttery textured, delicious, and costly. Bibb is sometimes called Kentucky limestone.

Though available daily, head lettuces suffer in quality from extreme heat or cold, so expect them to be less than perfect in midsummer or midwinter. Most of the hydroponic crop goes to restaurants, so it is hardest to find. Choose heads that are fresh and crisp. Avoid wilt, rot, and browning. Butterheads should be a healthy green and fairly heavy for their size, but hydroponic lettuce is always pale and lightweight. Iceberg is also usually rather pale, but it should feel quite heavy and solid. Wrap unwashed heads and refrigerate them for a day or two if necessary.

The undisputed king of the salad bowl, lettuce is combined with just about everything savory, as

well as with sweet fruits. And, of course, it is a popular sandwich ingredient. Head lettuce can be steamed, braised, or added to soups. Iceberg is not very exciting, but it does seem just right when finely shredded for Tex-Mex dishes. This is a clear violation of the general rule that lettuce be torn and not cut. Using a knife is supposed to produce bitterness and hasten browning. The bitterness is open to question, and browning should always be avoided by breaking up the lettuce, by whatever means, within a few hours of serving time. Iceberg is remarkably sturdy, but the others wilt instantly when they are dressed. Unless wilted salad is the aim, lettuce salads should be served immediately after dressing.

WHAT TO LOOK FOR:
crip heads of proper color and heft, very firm iceberg

WHAT TO AVOID:
wilt, rot, flabbiness, browning

HOW TO STORE:
wrap and refrigerate, unwashed, for a day or two

PRIME SEASON: (Indicated by darker shade).

| JAN | FEB | MAR | APR | MAY | JUN | JUL | AUG | SEP | OCT | NOV | DEC |

LEAF LETTUCE AND ROMAINE

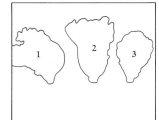

1. Red (bronze) leaf
2. Romaine
3. Salad bowl (green leaf)

Primitive is not the right word. Maybe *natural?* Leaf lettuces are most like their wild ancestors, untainted by the sophisticated heading trick taught to their cousins. And they taste good too. Substance, refreshing green flavor with a hint of bitterness–these are the qualities you can expect. The most popular of the leaf lettuces is the green one called salad bowl. Though developed in California, it has found a following in Europe as well. There are also the less crinkly ones called oak leaf, often tipped with red or bronze. Romaine, or cos, is not a true leaf lettuce, but rather a distinctive type and one of the most ancient. It is named for the Mediterranean island of Kos, home of Hippocrates, where it was extensively cultivated, and it was indeed very popular in Rome. Many historians believe that it first spread to western Europe with the popes who moved to Avignon. It should be mentioned that there is yet another type of lettuce, a rare one called stem lettuce, that is grown for its fat, crunchy main stem. This kind is most cultivated by the Chinese, and if you happened upon one, it would probably be called celtuce, asparagus lettuce, or *woo chu.*

Leaf lettuces and romaine are available year round. They stand up to summer heat better than head lettuces, but they too are less plentiful, and more costly, in midwinter. Look for crisp, green leaves. Avoid wilt, rust, and rot. These lettuces can be refrigerated, unwashed but wrapped against moisture loss, for a day or two if necessary.

Leaf lettuces and romaine are excellent salad and sandwich fare. They have the flavor and crunch to stand up to highly flavored sauces and additions. They are also fine for cooking–steaming, braising, and in soups. As with other lettuces, they are traditionally torn, rather than cut, except for the famous salad invented by Caesar at his restaurant in Tijuana. Without getting involved in the 'original' recipe for this garlic-, anchovy-, and cheese-flavored favorite, it is safe to say that romaine lettuce, its major ingredient, is usually cut into strips or squares.

WHAT TO LOOK FOR:
crisp, green leaves

WHAT TO AVOID:
wilt, rust, rot

HOW TO STORE:
wrap and refrigerate for a day or two if necessary

PRIME SEASON: (Indicated by darker shade).

JAN	FEB	MAR	APR	MAY	JUN	JUL	AUG	SEP	OCT	NOV	DEC

MÂCHE

Lamb's lettuce is one of its English names, which paints a charming picture of springtime and fields in which little lambs graze on velvety greens. This salad herb can sow itself in a variety of cultivated areas, which is why it is also called corn salad and field salad. Mâche is the usual French name, a word which has become popular recently in the U.S. along with the green itself, which was previously all but unknown to any but home gardeners and French or Italian immigrants. The French also call it *doucette*, 'little sweet,' *boursette*, 'little purse,' and *salade de chanoine*. Mâche is a very popular salad green in Europe, where it is widely cultivated but usually expensive. There is a wild variety in Italy called *insalatina*, 'little salad.'

Mâche is most prized in winter, when there used to be precious little in the way of salad greens. It is also available in fall and spring and maybe into summer, too. Choose very fresh blue-green leaves.

Avoid wilt, rot, and dryness. Mâche is very perishable, but it may be wrapped and refrigerated for a day or two if necessary.

Mâche is at its best in salads. The texture is pleasantly firm and chewy without being crunchy, and the flavor is bland and lettucelike. It can be combined with any of the usual ingredients for a tossed salad. Because of its cost, many cooks prefer to feature mâche as the main or single vegetable in a salad rather than burying it in a combination of many things. A vinaigrette made from finest quality wine vinegar and olive oil makes an ideal dressing. The pretty little leaves, usually spoon shaped, make an attractive presentation in salads that are carefully composed rather than tossed. Mâche can also be used as a potherb in any spinach recipe, where it would be wildly expensive and too neutral in flavor to be very interesting alone.

WHAT TO LOOK FOR:
very fresh blue-green leaves

WHAT TO AVOID:
wilt, rot, and dryness

HOW TO STORE:
wrap, refrigerate, and use within a day or two

PRIME SEASON: (Indicated by darker shade).

| JAN | FEB | MAR | APR | MAY | JUN | JUL | AUG | SEP | OCT | NOV | DEC |

MANGOES

The ultimate tropical fruit? Mangoes are prime contenders for the title. Their juicy sweetness and exotic perfume have earned them a wide following. India lost the monopoly long ago to tropics everywhere, and there is now a variety of shapes and sizes. Mangoes may be round like a peach, kidney shaped, pear shaped, or oval. Rounder varieties are often better choices because of the large central seed, which is very fibrous. In the less desirable varieties these fibers can invade the flesh, making it difficult to eat. Though the flavor of a ripe mango eludes easy description, the yellow to carrot-colored flesh reminds some people of a ripe peach with overtones of apricot, pineapple, and papaya.

Mangoes are available most of the year from supermarkets and greengrocers. The best crop comes to market in late spring and summer. In cold weather, mangoes from various parts of Central and South America are shipped north with varying degrees of success. Like other tropical fruits, mangoes are at their best only when tree-ripened, or nearly so, and they are highly perishable. Do not expect a perfect sweet mango in January, but you may find a surprisingly good one.

Choose plump fruit with smooth skin and a pleasant, fresh smell. A good mango will yield to gentle pressure. Skin color ranges from yellow to orange to red, or a combination of these, but a small amount of green is acceptable. Avoid very soft or withered fruit and bruises or large black spots. Small dark speckles, however, are not a problem in fruit that is otherwise in good condition. Very green or rock-hard mangoes have been picked prematurely and will never ripen. Fruit that is fresh-smelling, firm, and well colored can be nursed along at home in a brown paper bag at room temperature until it softens and loses most of its green color. Use immediately or hold, refrigerated, for a day or two.

Best known as the major ingredient in Indian chutneys, mangoes have a wide variety of uses. They are pickled and preserved, spiced or not, to create many relishes. Fresh, the mango is delicious (but messy) eaten out of hand or included in fruit salads, ice creams and ices, or in baking. Lime juice is a classic flavoring. Mango is a fine complement to duck and can be used underripe in stews or ripe and sweet with breast of duck in artfully arranged salads.

It is always best to serve fresh mango very cold. Chilling helps to disguise the slight odor and taste of turpentine present in some varieties. The skin has oils that produce allergic reactions in some people, so it is a good idea to consider serving only peeled mango to guests.

An ancient preparation for mango involves slicing off the flat sides close to the pit, deeply scoring the flesh into a crosshatch, and reversing the arc of the skin to push the flesh up into rows of standing cubes. The result is dramatic and flowerlike.

WHAT TO LOOK FOR:
plump fruit, smooth skin that has at least begun to color, slight softness, fresh smell

WHAT TO AVOID:
very green or rock-hard fruit, shrivel, mushiness, bruises, rot, large black spots

HOW TO STORE:
refrigerate only when completely ripened, for a day or two

PRIME SEASON: (Indicated by darker shade).

| JAN | FEB | MAR | APR | MAY | JUN | JUL | AUG | SEP | OCT | NOV | DEC |

MELONS

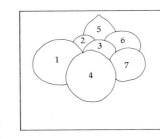

1. Santa Claus – so named because they are rumored to last until Christmas, rind green and gold but not too green, moderately fragrant, stem cut.

2. Persian – larger and rounder than cantaloupe (U.S.), often greener but should be on the way to golden, fragrant, stem cut.

3. Orange honeydew – looks like a lightly netted honeydew but soft orange in color, moderately fragrant, stem out.

4. Casaba – little or no green, not very fragrant, stem cut.

5. Crenshaw – little or no green, except in September, fragrant, stem cut.

6. Honeydew – rind creamy white and sticky, not very fragrant, stem cut.

7. Cantaloupe (U.S.) – skin heavily netted and golden beneath, very fragrant, no cut stem but instead a smooth scar.

Melons are that rare combination, that exception to the rule – they taste sweet and satisfying but are remarkably low in calories. And they even have the bonus of measurable amounts of vitamins A and C. Not a bad package. Sweet melons have been great favorites for centuries, long before anyone thought of dieting. They have an exotic lushness and perfume that has made them international stars.

Nomenclature is a bit of a problem here. Cantaloupes are very popular in Europe, named for the papal gardens of Cantalupo near Tivoli. They have green or orange flesh and skin that is smooth, scaly, wrinkled, or ridged, but never netted. The netted ones are muskmelons, and that's what the favorite American melon is, the one called cantaloupe even though no true cantaloupes are grown commercially in the U.S. Persians and American cantaloupes are the common muskmelons, and you might hear them referred to as cassia or nutmeg melons, a reference to their rich, spicy scent. American cantaloupes have the distinction of separating from the vine when they are ripe, leaving a smooth scar on the stem end, a feature to look for. The others – casaba, honeydew, crenshaw, Santa Claus, and the like – are usually called winter melons because they keep well. These are likely to be picked young and must be tested carefully. All of these melons freely interbreed to produce new varieties.

Choosing a perfect melon requires some knowledge. Here are the basic rules:

Melons should be fragrant. The winter melons are iffy here. They can still be acceptable with little or no odor.

The stem end should yield to gentle pressure. This test is not wholly reliable because of all the other shoppers who may be testing and bruising the fruit.

Skin color should be losing its green. Scaly irregularities in the skin are not a quality factor.

Melons should feel heavy, very heavy.

Avoid soft or watery spots, shriveled skin, overall flabbiness, cracks, or rot. Melons do not gain sugar after picking, but they may improve in juiciness and softness with a few days at room temperature. Refrigerate closely wrapped to avoid perfuming everything else in the refrigerator, especially the butter. Use promptly.

Melons are most often eaten alone for breakfast or dessert, perhaps with a squirt of lemon or lime juice or, strange to tell, salt and pepper. They are good in combination with one another and any fruit for salads, where they might be flavored with honey, rum, or mint. Any good ripe melon is delicious with thinly sliced prosciutto or other fine raw-cured ham for hors d'oeuvres. Melons are sometimes teamed with cold shrimp or crab for warm-weather luncheon entrées.

WHAT TO LOOK FOR:
fragrance (in some varieties), stem end yielding to gentle pressure, proper color, heaviness

WHAT TO AVOID:
soft or watery spots, shriveled skin, overall flabbiness, cracks, rot

HOW TO STORE:
age for a few days at room temperature if desired, then refrigerate, well wrapped, and use promptly

PRIME SEASON: (Indicated by darker shade).

| JAN | FEB | MAR | APR | MAY | JUN | JUL | AUG | SEP | OCT | NOV | DEC |

SPECIAL MELONS

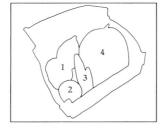

1. Spanish melon
2. Israeli muskmelon
3. Chinese bitter melon
4. Chinese winter melon

There's melons and then there's melons. Some is and some ain't. In addition to the sweet summer and fall melons (now available many months of the year), there are other crops around the world that range from the local to the seasonal to the just plain ethnic. And some are sweet melons and some are squashes that are called melons anyway. All warm climates can produce fruits of either sort. Among the true melons for export is the Israeli version of cantaloupe (U.S.), a small November and December thing that can be quite good. It is green fleshed, but unmistakably a muskmelon. Spanish melons from Spain or Sicily are usually less successful, the victims of the familiar shipping problems that dictate premature harvest. Among the phony melons are two famous Chinese ones that are available all year long. Bitter melon is a summer squash that is otherwise aptly named for its refreshing menthol flavor. Winter melon is a winter squash whose name has nothing to do with the season, but everything to do with its thick waxy bloom, which suggested hoarfrost to the ancients and still does today.

True sweet melons should be heavy and firm without any rot or soft spots. The blossom end should yield to gentle pressure, and a pleasant fragrance is an indication that there will be sweetness and flavor within. Israeli melons should have lost their greenness and become golden between the webbing. Do not refrigerate unless they are very ripe and fragrant. Bitter melon and winter melon should be firm and unblemished. Bitter melons can be wrapped and refrigerated for a day or two; so can winter melon, but it is usually sold cut in pieces and therefore will turn mushy if kept too long.

Sweet melons have the usual uses, while Chinese 'melons' are for cooking. Bitter melons are seeded and sliced, parboiled, and then used in a variety of dishes–stir-fries, braises, and soups. The flesh of winter melon is most often cut into pieces and simmered in a soup made with pork or ham, where its slight crunch is highly prized. The flavor itself is quite bland and is always used in combination with savory ingredients. For special occasions, the whole squash is hollowed out, the frosty bloom rubbed away, the deep-green rind decoratively carved to create a beautiful container for serving the soup. Winter melon pieces can also be stir-fried or even candied.

WHAT TO LOOK FOR:
firm, heavy melons of proper color; sweet melons must be fragrant

WHAT TO AVOID:
rot, bruises, mold, mushiness

HOW TO STORE:
refrigerate for a day or two; sweet melons should be chilled only if very ripe and then carefully wrapped to avoid sharing the scent with everything else in the refrigerator

PRIME SEASON: (Indicated by darker shade).

	JAN	FEB	MAR	APR	MAY	JUN	JUL	AUG	SEP	OCT	NOV	DEC
Chinese melons												
Sweet melons												

CULTIVATED MUSHROOMS

In the history of food, mushrooms take a prize for fear and controversy. There are two very different reasons for this situation. First, choose the wrong one and you die a sudden death, die a slower death with agony, or just pray for the former. Second, mushrooms grow mysteriously, popping up suddenly overnight where no visible growth was before, changing shape rapidly and then collapsing into powder or ooze; in short, the work of the devil. Some varieties will invade a field and grow in large rings, devil's rings, which used to send the fearful right out to find the nearest holy man. Nevertheless, mushrooms of various sorts have been cultivated for a long time, completely controlled and reliable since the nineteenth century.

Cultivated mushrooms are available in steady supply all year. Some are very smooth and white, while others can be off-white or tan and sometimes slightly scaly, depending upon the variety. Size is a factor in relation to intended use, but not to quality. Choose firm, fresh-looking mushrooms that are reasonably clean with short stems (for economy) and tightly rounded caps. Once the cap begins to spread and breaks the 'veil' that joins it to the stem, the mushrooms are old and should be purchased only at reduced price. Avoid bruising, rot, moisture, and mushrooms that have been washed and treated with bisulfites–they taste awful and the chemicals are probably poisonous. If the treated ones are the only mushrooms available, they can be washed in hot water to remove some of the taint, but the tex-ture will suffer. Fresh mushrooms can be stored, loosely wrapped and refrigerated, for a day or two if necessary.

Cultivated mushrooms are bland, but they do add a pleasant texture, and a little flavor, to a wide variety of savory dishes. They can be eaten raw, stewed, fried, sautéed, baked, broiled, stuffed, marinated, pickled, and used in salads, soups, or sauces. They go with any foodstuffs that are not sweet. Very fresh mushrooms and brief cooking are the rules, and the tender caps are considered the choice part. Older specimens and stems can be used to flavor sauces, or they can be finely chopped and sautéed in butter with minced shallots and Madeira to produce *duxelle*, a versatile paste for coating, stuffing, or flavoring other foods. Unless they are to be cooked in liquid, mushrooms should not be washed, and they should never be soaked–they will sponge up amazing quantities of liquid, water, or fat. The alternative cleaning method is simply to remove the soil with a very soft brush or damp cloth. Peeling of cultivated mushrooms is a silly waste of both time and mushroom.

WHAT TO LOOK FOR:
plump, round, firm mushrooms, short stems

WHAT TO AVOID:
open caps with gills showing, bruises, rot, discoloration

HOW TO STORE:
wrap loosely and refrigerate for a day or two if necessary

PRIME SEASON: (Indicated by darker shade).

| JAN | FEB | MAR | APR | MAY | JUN | JUL | AUG | SEP | OCT | NOV | DEC |

Wild mushrooms

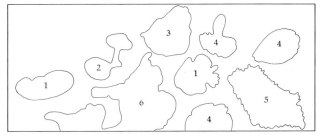

1. Oyster mushroom (pleurotte)
2. Shitake
3. Lobster mushroom
4. Large Boletus
5. Enoki
6. Chanterelle

A biologist once recommended these steps for the amateur mushroom gatherer:

1. Go hunting only with an expert mycologist, of which there are just a few in the world.
2. Gather only the varieties that he gathers.
3. Refrigerate your harvest until he has eaten his.
4. If he lives, then enjoy your mushrooms.

Though extreme, this advice is not without merit, for the deadly toadstool has power to assume a pleasing shape, mimicking the edible fungi. Fortunately, some of the wild flavorful mushrooms can be cultivated. Others are gathered by experts who build their business on the safety of the product.

BOLETUS

The most usual of the wild mushrooms are several varieties that are known collectively as boletus. These include *cèpes* in France, *porcini* in Italy, and *shitake*, originally from Japan, that come in various sizes from normal to large, called golden oak, to huge. All boletuses feed on rotting wood, often oak, and these conditions have been duplicated by growers with some success, at least for shitakes. Shitakes tend to have tough fibrous stems that cannot be tenderized (use for flavor, do not discard), while the others have softer, edible stems. The rich, woodsy flavor is delicious in a variety of cooked dishes–soups, sauces, stews, casseroles, stir-fries, vegetable combinations, and a simple broil or sauté.

CHANTERELLES

These have a very showy orange color and trumpet shape. There are large ones and smaller ones called *girolles.* Evergreen forests are the natural habitat, and some advances are being made with cultivation techniques. Chanterelles are usually stewed.

ENOKI

This pretty little Japanese mushroom takes well to cultivation and is now becoming widely available on a year-round basis. The flavor is rather bland, but it adds a nice crisp note to stir-fried dishes and salads.

LOBSTER MUSHROOMS

This big weird fungus is indeed lobster-red. It is best when sliced and gently sautéed in butter or olive oil.

OYSTER MUSHROOMS

So named for their soft grey color and shape, these are also called *pleurottes.* Poplar is the wood they like, and cultivation is under way. Oyster mushrooms have a very firm, meaty texture that complements elegant dishes like braised or sautéed loin of veal, or earthy food like roast loin of pork, sausages, or salt cod. They go well with game.

MORELS

Perhaps the rarest and costliest of the wild fungi (after truffles) are the morels. They have distinctive, wrinkled, conical caps, and are wonderfully earthy and smoky. They cannot be cultivated and are extremely hard to come by fresh. They are most likely to be found dried, a state that is, if not ideal, still delicious. Morels are most often cooked with cream.

PRIME SEASON: (Indicated by darker shade).

| JAN | FEB | MAR | APR | MAY | JUN | JUL | AUG | SEP | OCT | NOV | DEC |

WHAT TO LOOK FOR:
firm, fresh-looking specimens

WHAT TO AVOID:
wilt, bruising, rot, moisture

HOW TO STORE:
refrigerate, loosely wrapped,
for a day if necessary

Nuts

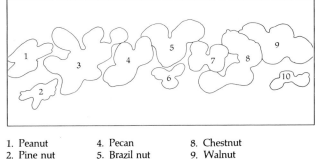

1. Peanut
2. Pine nut
3. Almond
4. Pecan
5. Brazil nut
6. Ginkgo
7. Hazelnut
8. Chestnut
9. Walnut
10. Pistachio

Nature did one of her best packaging jobs on nuts, the edible seed kernels of various plants. So safely encapsulated are the delectable innards that they keep for months at room temperature. In fact, the shells not only discourage decay, but also animals without the right dentition, including man. Most consumers prefer the results of trained laborers and machinery to their own efforts with a nutcracker, and shelled nuts that are vacuum-packed–the only reliable storage system–can indeed retain some freshness. But for maximum flavor and for the sheer nostalgia of it all, there is nothing quite like the real, whole nut. Most of them are harvested in fall and should remain in good condition into winter. We will not bog down in scientific concerns about what is and is not a true nut, but rather follow common usage. Consumers should know that most nuts have high levels of natural oils and are quite caloric.

Almonds

The most important Old World nut is also very popular in the New. It is extremely versatile in savory and sweet dishes, whole, sliced and chopped, ground into flour, and pressed for oil. Almonds are called sweet almonds to distinguish them from bitter almonds, the source of all almond flavorings, which are not almonds at all but apricot kernels.

Brazil nuts

Surprise! They really are native to South America. These fat, hard-shelled nuts are delicious and very oily, and are therefore among the most perishable.

Chestnuts

This is the only true nut with appreciable starch levels, making it useful as a vegetable as well as for many sweet preparations. It is also perishable. The famous chestnut blight of 1904 wiped out every American chestnut by 1940, so now there are only the superior ones of Italy and France and the mediocre ones from Asia.

Ginkgo nuts

For ornamental purposes, only male ginkgo trees are grown, because the female gives off an offensive smell when it is fruiting. But there are hardy souls who grow them to supply nuts for Asian markets. Ginkgo nuts have an odd sweetish flavor and gummy texture (by Western standards) and they are prized for Chinese soups and some poultry and vegetarian dishes.

Hazelnuts

Filberts and hazelnuts are probably different, the former Old World and the latter New. But the two terms have become synonymous for this elegant nut that is so deliciously used, especially in France and Italy, for cakes, cookies, chocolates, ice cream, oil, salads, and butter.

Peanuts

This all-American favorite is also popular in Africa, China, and Southeast Asia, and it is one of the world's most important food crops. The name is apt, if inaccurate, for while it behaves like a nut, it is a pea of sorts–a legume. Peanuts have the curious habit of growing underground–the flower stalks bend down and grow into the earth, where the peanuts can safely mature. In versatility and nutritional value, they rival soybeans.

Prime season: (Indicated by darker shade).

JAN	FEB	MAR	APR	MAY	JUN	JUL	AUG	SEP	OCT	NOV	DEC

PECANS

This American relative of the walnut is highly prized, especially in southern cooking, for its rich, almost mapley flavor. Though it is most often used in pies, cakes, cookies, and candies, it has become respected around the world for its versatility in the kitchen.

PINE NUTS

The sweet little kernel of various Mediterranean pines lends its pleasant slight piney flavor to a variety of preparations from the famous *pesto alla genovese* to stuffed vine leaves and Arab meat loaves to sweets. Pine nuts, or *pignoli,* are expensive, perishable, and usually only available husked from the cone.

PISTACHIOS

The lovely color called pistachio green is but one reason for the popularity of these Middle Eastern natives. The flavor is subtle and delectable, equally at home in savory and sweet dishes, especially *charcuterie* (sausages, terrines, pâtés, and so on), pastries, and ice cream.

WALNUTS

The major walnut is the blond one called English. No nut is more versatile. Beyond munching and many sweet and savory preparations, walnuts are also pressed for oil and even pickled.

The thought of rarer nuts brings to mind hickories, which nourished American Indians for centuries, and the black walnut, definitely a tough nut to crack. Only people with a tree in the backyard want to bother with these, but it is possible to find shelled black walnuts. They lend an extraordinarily rich flavor to baked goods. Whole cashews are sometimes available in Asian markets. Though they seem like famine food, especially to the children who are the major consumers these days (after squirrels, of course), acorns were a major food source in premodern times. And humans eat, or have eaten, butternuts, beechnuts, the small American relative of the chestnut called chinquapin, and dozens of other local nuts.

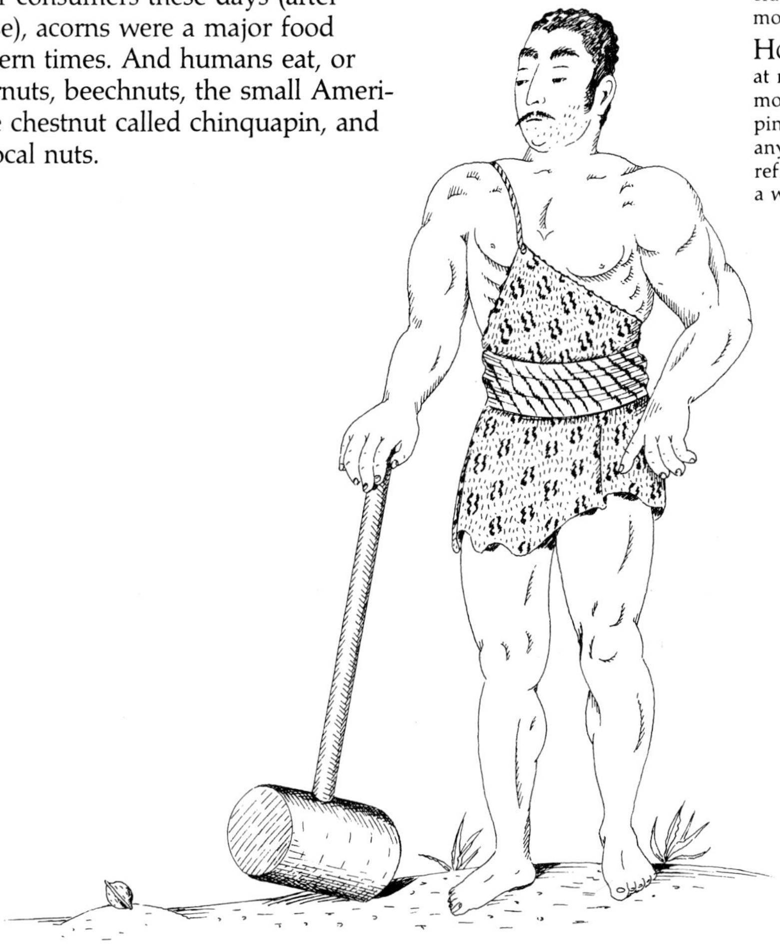

WHAT TO LOOK FOR:
sturdy whole shells (pistachios will be open) with no rattle when shaken (except for peanuts)

WHAT TO AVOID:
cracked shells, worm holes, moisture, rancid odor

HOW TO STORE:
at room temperature for a month or more; pistachios, pinenuts, and chestnuts or any shelled nuts should be refrigerated and used within a week or so

OKRA

People are so funny about unfamiliar foods. Okra is one of those that gets a long face from the uninitiated. 'Ugh, slimy,' they say, or something similar. Would a different name help? In Britain okra is also called ladyfingers and it still hasn't caught on. How about hibiscus? That's what it is. This African native is staple fare from the Mediterranean to India and in the American South, but elsewhere the citizenry needs a bit of prodding if we intend to make this delicious vegetable a favorite.

Fresh okra is available all year with the largest volume in summer and the smallest in midwinter. There are considerable differences in the varieties that come to market. They range not only in size, but also from bright to dull or pale green and from hairy to smooth. Overgrown pods are likely to be woody, so the best tactic is to handpick the smallest of the batch, under three inches long if possible. Okra is quite perishable, so care should be taken to choose only the plump green ones. Rust sets in quickly, and although a very small amount of darkening around the edges does no great harm, do avoid okra that is turning black, shriveled, or soft. Use immediately if possible, or refrigerate, wrapped closely, for one day.

The smooth green flavor and mucilaginous texture of fresh okra make it a satisfying vegetable when simply steamed to tenderness and served with butter. But it is more often combined with other things, especially tomatoes, corn, peppers, and onions. The famous soup/stews of Louisiana, gumbos, got their name from one of the African words for this plant, which owes its foothold in the New World to the slave trade. Gumbos are always thickened with okra or gumbo filé powder made from sassafras leaves, or, horror of horrors for purists, both. Okra is delicious when breaded with cornmeal and deep-fried, and when pickled. In Arab and Indian cooking it combines well with other vegetables and with meats in spicy preparations. Okra freezes more successfully than most vegetables, and frozen okra is an acceptable substitute for fresh in dishes with a lot of ingredients and fairly long, slow cooking.

WHAT TO LOOK FOR:
plump, green pods, preferably under three inches long

WHAT TO AVOID:
rust, blackening, shrivel, softness, overgrown pods

HOW TO STORE:
wrap closely and refrigerate for a day or two

PRIME SEASON: (Indicated by darker shade).

| JAN | FEB | MAR | APR | MAY | JUN | JUL | AUG | SEP | OCT | NOV | DEC |

ONIONS

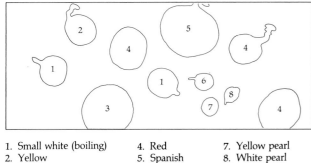

1. Small white (boiling)
2. Yellow
3. White Bermuda
4. Red
5. Spanish
6. Red pearl
7. Yellow pearl
8. White pearl

Oh, yes. Tears have been shed. And it is not the beauty of the lilies that prompts the tears, though onions are indeed lilies, along with shallots, leeks, chives, scallions, and garlic. Of all the kitchen lilies, onions are by far the most popular, used in some form in every developed cuisine in the world. They contain sulfur compounds and volatile oils that delight the palate, perfume the air, and cause extraordinary eye irritation for those who are sensitive to them. There probably exists more old wives' hocus-pocus for cutting onions than for any other kitchen operation. Cooks have been instructed to dissipate the fumes by working near an open flame, or near or under running water, or to trap the fumes before they reach the eyes by holding a piece of bread in the mouth. Chilling the onions before cutting them has been suggested, along with a variety of other tricks of varying sensibleness.

As usual, the best way is the right way. Onions should be cut with a very sharp knife properly wielded–a sawing motion that cuts rather than bruises. Airtight goggles are the solution for the supersensitive. Also, cooks who use carbon steel knives must wipe them frequently to avoid staining the knives and the onions. Gadgets and food processors generally do a bruising job equal to or worse than poor knife technique, and they should be avoided. Lingering odors on hands and equipment can be reduced by rubbing with salt, lemon juice, or vinegar.

Onions are plentiful all year with higher quality in cold weather. Choose onions that are firm, heavy, and unblemished, with dry, papery skins and thin, dry, pliable stems that have no tough core. Avoid softness, sprouting, moisture, rot, or black mold. Depending upon the variety, onions should keep well for a week or two, or even months in winter if they are stored in a cool, dry, dark, well-ventilated place. Lacking such ideal conditions, you may store them anywhere at room temperature for use within the week. However they are stored, it is important that onions not be kept in airtight packaging, for moisture encourages decay. Refrigerators are too humid. Should onions begin to sprout, you may use the sprouts like scallions and the bulb can still be used if it is in good condition.

Onions are eaten raw and used in any savory cooking technique, and they can be made into sweet relishes and even baked into pies. There are a few basic rules that should be noted. When onions are boiled, they should actually be simmered gently for the most delicate flavor. Simmering and steaming produce the least residual odor on the breath. Onions to be cooked whole should be deeply pierced at the root end with a knife in a cross pattern or with a skewer to release pressure and minimize exploding centers. For cooking in fat, the shallower the fat, the shallower the layer of onions should be, so that they actually cook in the fat and do not steam themselves into a sulfurous state. High heat and vigorous stirring will get them off to a good start, but the flame must be adjusted so that they do not burn. Burning produces acridity. Minced or sliced onions can also be gently stewed in a good amount of butter or oil to create a delicate flavor. A preliminary partial cooking in fat is standard practice when onions are to be added to stews and the like, and it improves their flavor.

PRIME SEASON: (Indicated by darker shade).

| JAN | FEB | MAR | APR | MAY | JUN | JUL | AUG | SEP | OCT | NOV | DEC |

With care, onions can be gently stewed, alternately covering the pan to sweat out their moisture and uncovering it to boil the moisture away, until the onions have reduced considerably in volume, caramelized, and concentrated their natural sugar to a remarkable sweetness.

Considering that onions are so common, it is amazing how little exact information exists for consumers concerning the names of the types and their properties. Confusion abounds because of differences in local and individual nomenclature, and the fact that the same variety of onion can vary with growing conditions–witness the Vidalia onion that came to prominence during the tenure of President Jimmy Carter: it is an ordinary yellow onion that develops an extraordinary sweetness when grown in a particular part of Georgia. With no attempt at botanical accuracy, here are the basic onion types:

PEARL ONIONS
These are tiny gems that usually come in white, but may be yellow or red as well. Pearl onions are peeled and pickled, or boiled, braised, or creamed for vegetable combinations or stews.

YELLOW ONIONS
These are medium-sized, multipurpose cooking onions that are usually firm fleshed and highly flavored. The globe-shaped ones are best. In the spring, southern onions come on the market; these are the flat-topped or *granex* onions that form most of the supply through the summer until the northern crop has been harvested and cured in fall. Southern onions tend to be mild, moist, and soft textured. They do not hold their shape well when sautéed.

SPANISH
It is one of the clearest terms, generally applied to large, globe-shaped, yellow onions that are mild and sweet. There are large white Spanish onions too, though they will usually not be so named in the market. Spanish onions are excellent for eating raw and can also be cooked.

WHITE ONIONS
They come in a variety of shapes, sizes, and tastes. There are small ones, often called boiling onions, that have the same uses as pearls. There are medium-sized white onions, and there are large, mild white onions sometimes called Bermuda, one of the most confusing of all onion terms. The large ones are for eating raw and for fried onion rings.

RED ONIONS
They also have a variety of sizes, shapes, flavors, and names. They are often mild and eaten raw, but only tasting will tell. For cooking, red onions should be used advisedly because of their color. As if to cause confusion, the large mild ones are often called Bermuda, or sometimes Italian, though that name is more accurately reserved for the medium-sized elongated red onions.

WHAT TO LOOK FOR:
firm, heavy, unblemished onions; dry papery skin; thin, dry, pliable stems with no tough core

WHAT TO AVOID:
softness, sprouting, moisture, rot, or black mold

HOW TO STORE:
cool, dry, dark, well-ventilated place (root cellar) for up to a few months, or room temperature, ventilated, for a week

MANDARIN ORANGES

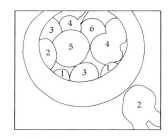

1. Tangelo
2. Tangerine
3. Temple orange
4. Mineola tangelo
5. Clementine
6. Mandarin

Given that oranges are Chinese or Southeast Asian natives, mandarin is not an unusual name. Legend has it that these fruits were named for their resemblance, in shape and color, to the buttons atop the hats worn by those officials of the Chinese empire. With that charming thought in mind, we will endeavor to make some sense of the modern types and their various names. Tangerines are popular fruits in the West, named for the city of Tangiers in Morocco that was long famous for them. Tangerines are the rather flattened ones with loose skin and segments that come apart easily. Tangerines have been hybridized with other citrus to produce various 'tangors,' most notably temple oranges, which are believed to be a sweet orange and tangerine cross. Another important hybrid is the result of tangerine and pummelo or grapefruit cross-pollination, the tangelo. Tangelos are the largest of the group, deeply colored, and ranging in shape from flattened like a large tangerine to distinctively knobbed at the stem end when they are of the Mineola type. Clementines are the babies of the lot. They are probably a tangerine/orange hybrid and are sometimes seedless. As if to encourage more confusion, there is also a round fruit of tangerine size that is simply called mandarin.

The mandarin oranges are cold-weather fruits. The first ones ripen in the fall, but you will not find them all until tangelos, temples, and honey tangerines arrive in December or January. Midwinter is peak season, but you may find some of them well into spring. Choose plump, bright fruit that is tight skinned and heavy for its size. Tangerines are the only ones that tend to be a bit baggy, but even they should not be too lightweight or loose skinned. Beyond these tests, only tasting will reveal quality. It is a good idea to buy one fruit and try it out. Each of the mandarins has the potential for being delicious or boring. Be especially careful to avoid dry, spongy flesh that results from chill damage, a far too common flaw. The mandarin oranges should keep well for a week or so if kept cool, preferably in the warmer vegetable section of the refrigerator.

The mandarins are most often peeled and eaten out of hand. They do lend a festive air to winter meals and snacks. They can also be used in salads and desserts or cooked to flavor poultry or seafood preparations. Seedless clementines are delicious dipped in bittersweet chocolate. Tangerine skin may be dried in a slow oven to replace the commercial product used in Chinese stir-fried dishes, especially orange beef.

WHAT TO LOOK FOR:
plump, bright, tight-skinned, heavy fruit (tangerines will be a little loose)

WHAT TO AVOID:
dull, shriveled, bruised, soft, loose-skinned, lightweight fruit, mold

HOW TO STORE:
chill, for a week or so

PRIME SEASON: (Indicated by darker shade).

| JAN | FEB | MAR | APR | MAY | JUN | JUL | AUG | SEP | OCT | NOV | DEC |

SEVILLE ORANGES

Depending on whom you believe, sour oranges may be the original ones from which all the others were bred. If so, the parent is not being honored much these days. Seville, or bitter orange, is the sour orange most grown, but the harvest is tiny compared with the volume of sweet oranges and mandarins. There are even reference works indicating that Seville oranges are not edible. A vicious lie. What they are not is sweet. Anyone who ever tasted English marmalade knows that they are not only edible but delicious. And only this fruit will make a perfect *sauce bigarade* for duckling. And those who fancy orange-flavored liqueurs have been enjoying the flavor that comes from the essential oils in the rind of Seville oranges. The crop destined for fresh market is absurdly small. Consumers and suppliers have taught each other that we can do without fresh Seville oranges, and it is up to the public to demand more of them.

Seville oranges have a short season. If you can find them at all, it will be in February or March. They are not very pretty fruits, for no attempt has been made to standardize and cosmeticize them for mass appeal. As with other oranges, they should be plump, firm, and heavy for their size, which varies. Skin color will normally range from orange to yellowish and maybe greenish. A small amount of russeting is acceptable. Avoid soft, bruised, or shriveled fruit and, definitely, avoid mold. Seville oranges in good condition will keep for a week or better if they are bagged and refrigerated.

Homemade Seville orange marmalade is one of the most sublime of all confitures. Purists insist on nothing but oranges and sugar, but adventurous cooks might want to try fruit mixtures. The pectin and acid levels of Seville oranges are ideal for preserving. Orange-flavored desserts get a lift from these oranges, and the zest (peel) can be used to flavor a variety of things, including custom-made liqueurs. Seville oranges are sadly underused outside French kitchens. Their sprightly tartness is just right for sauces to enhance fatty meats – duck, goose, and pork – and for game. The juice makes a delicious vinaigrette, especially with walnut or hazelnut oil.

WHAT TO LOOK FOR:
plump, firm, heavy fruit
WHAT TO AVOID:
softness, bruising, shrivel, mold
HOW TO STORE:
bag and refrigerate for a week or more

PRIME SEASON: (Indicated by darker shade).

| JAN | FEB | MAR | APR | MAY | JUN | JUL | AUG | SEP | OCT | NOV | DEC |

Sweet Oranges

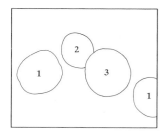

1. Blood oranges (Moros)
2. Valencia
3. Washington navel

These blood oranges are of the sort with deep red flesh and reddened skin, while others look like orange oranges until they are cut.

Many historians believe that sweet oranges are the original ones, coming from southern China and vicinity. Oranges were cultivated there for centuries in prehistoric times; then they followed the trade routes west to the Arabs, who in turn took them to northern Africa, Italy, and Spain. By the sixteenth century oranges not only had changed eating habits in Europe, but had led to increasingly lavish greenhouses, called *orangeries*, that for centuries remained one of the most important status symbols of the privileged. Spanish explorers and missionaries established their precious gift in the Americas, and the rest is history.

The major types of sweet oranges (not including mandarins) are juice oranges, navels, and blood oranges. The most widely grown of the juice oranges is Valencia, a type that is tasty enough for the table too. Others fill in the seasons, including a fine sweet late-winter orange from Israel called Jaffa. Navel oranges originated in Brazil and are now important in California and other growing areas. The Brazilian oranges made a stopover in D.C., so they came to be called *Washington navels*. They have excellent flavor with a fine sweet/tart balance. Blood oranges are an important crop in Italy, where they range from tart to wonderfully sweet. They also range in color, from orange lightly shot with red to so red that the rind is tinted red too. Oranges are available all year, but they are at their best in winter and spring. Government regulations ensure maturity at picking time (oranges cannot improve off the tree), so consumers need only look to see that oranges are in good condition–plump,

firm, heavy, free of mold or bruises. Skin color is a result of the variety and the climate. Color is not a quality factor, but color-added fruit should be avoided on general principle. Oranges keep well a week or two when wrapped and refrigerated.

Beyond their excellence juiced or eaten plain, oranges are fine salad fare combined with other fruits, or with onions, chiles, spices and herbs, vinaigrette, and the like. They can be used to flavor poultry, seafood, and many desserts. *Sauce maltaise* is a classic variation on *hollandaise* that is flavored with the juice and zest of tart blood orange. Orange zest is an important flavoring or garnishing agent, sometimes cut into fine *julienne* strips and candied. When removing the zest, be sure to include only the thin colored part where the oils are. The white pith is bitter and never used. A swivel-blade peeler is the best tool.

WHAT TO LOOK FOR: plump, firm, heavy fruit; thin skin

WHAT TO AVOID: softness, irregular shape, bruises, mold

HOW TO STORE: bag and refrigerate for a week or two

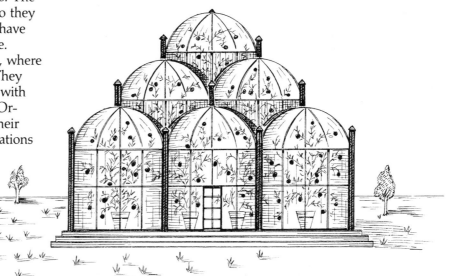

PRIME SEASON: (Indicated by darker shade).

| JAN | FEB | MAR | APR | MAY | JUN | JUL | AUG | SEP | OCT | NOV | DEC |

Papayas

Only the middle papaya is an excellent choice. The front one may ripen nicely provided it is very firm, and the two green ones will not ripen to sweetness and can only be used for cooking.

One of the best-kept secrets around is just what people are doing with all the papayas that are being shipped from the tropics to temperate regions. Most of these are hopelessly immature for eating as sweet fruit, yet they sell. Green papayas can be baked like squash or preserved, yet there is no evidence that this practice is widespread in Europe or the U.S. Green papayas (and their leaves) are rich in papain, a powerful enzyme that digests protein, yet outside the tropics the only commercial products containing papain seem to be in common use as meat tenderizers or digestives. So unless there is some mysterious use as yet unreported, people must be eating all those green papayas, and one wonders why. Of course, tasty sweet papayas get through too, but they are hardly in the majority.

Though papayas are in regular supply, sweet ones are only pleasant surprises and not to be counted on. In theory, a papaya with solid green skin will not improve, but one with a fair amount of yellow can be ripened in a paper bag with a few holes in it at room temperature. In practice, only green-and-yellow papayas that are quite firm stand a chance of improving. Soft green-and-yellow pa-

payas started out as the solid green ones that will not ripen once picked. So choose carefully, selecting firm fruit with the most yellow to the skin. Be sure to avoid blemishes or pebbly skin, for rot spreads quickly and dehydration is common. In fact, you might want to add humidity to the ripening bag by including a small vessel of water. A ripening time of several days is to be expected, and papayas are ready to eat when they are soft to the touch, but by no means mushy, and the skin has turned almost uniformly yellow-gold. They should be used promptly or wrapped and refrigerated for a day or two.

Sweet papayas make delicious breakfast or dessert fruits, alone with perhaps a squirt of lime juice, or combined with other fruits for salads or elaborate preparations. Papaya purée could be used for beverages, ices and ice cream, mousses, and so on. Papaya halves may be stuffed with shellfish or chicken salads for luncheon main courses. Their nickname, tree melon, comes from the physical appearance of the fruit, and while the flavor is not really melonlike (being more exotic and tropical), like melon, papaya goes well with fine raw-cured ham, prosciutto, for antipasto. Some people like to munch on a few of the seeds, but others will find papaya seeds too pungent.

WHAT TO LOOK FOR: firm, unblemished fruit that is green and yellow

WHAT TO AVOID: solid green or mushy fruit, bruises or rotten spots, pebbly skin

HOW TO STORE: ripen in a paper bag at room temperature until softened and golden, then refrigerate for a day or two if necessary

PRIME SEASON: (Indicated by darker shade).

| JAN | FEB | MAR | APR | MAY | JUN | JUL | AUG | SEP | OCT | NOV | DEC |

Parsley

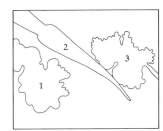

1. Flat-leafed (Italian)
2. Parsley root
3. Curly

If we had to choose a green herb, just one, to live with day in and day out, to provide seasoning and salad needs, parsley would be the logical choice. In fact, it has the essence of fresh green herbal flavor–no more, no less. Other herbs start with what parsley has and add on to it with results that are sometimes subtle and sometimes extravagant. But parsley remains pure and simple. It is easy to praise for its many virtues. Not only does it taste good, but it is remarkably nutritious–low in calories and rich in vitamins C and A, potassium, iron, folic acid, and calcium. It is versatile, available, and relatively inexpensive. Parsley looks pretty and is reputed to freshen breath after onions and garlic. The ancient Greeks and Romans favored parsley for festive occasions, not for its appearance or flavor but because garlands worn around the head or neck were said to protect the wearer from drunkenness.

There are three types of parsley in common use–curly, flat-leafed or Italian, and parsley root, also called Hamburg or turnip-rooted parsley. The leafy ones are plentiful all year with slightly smaller volume in winter and spring. Flat-leafed parsley has considerably more flavor than the curly variety. Choose fresh dark-green leaves. Avoid wilt, yellowing, and watery bruising. Parsley root should be plump and crisp with fresh leaves attached. Parsley needs cold refrigerator temperatures and high humidity. Some people like to refrigerate it with the stems in a glass of water and the leaves wrapped to protect them from moisture loss,

but there is no proof that this method is more effective than creating a humid atmosphere inside a plastic bag or wrapper. When properly stored, fresh parsley should keep nicely for a few days.

Beyond its use as a garnish, parsley is delicious in soups, salads, sauces, stews, and casseroles, and for flavoring all stocks. It is the most important of the *fines herbes* (usually said to include chervil, tarragon, and chives), and it is always included in an herb bouquet or *bouquet garni*, but only the stems are used so that the leaves may be saved for eating. Also, any green leaves will tend to turn a white stock grey. Parsley goes with all meats, poultry, and seafood, in addition to eggs, cheeses, compound butters, and other vegetables, especially beans, potatoes, onions, shallots, and garlic. Deep-fried parsley is an elegant edible garnish, and parsley can be made into jelly. To mince fresh parsley, make sure that it is as dry as possible and use a sharp knife. Parsley root can be peeled and cut up for soups, or it can be used like celery root in salads and cooked preparations.

WHAT TO LOOK FOR:
fresh, dark-green leaves;
plump, crisp roots

WHAT TO AVOID:
wilt, yellowing, watery bruising, shriveled or soft stems

HOW TO STORE:
moisten slightly, wrap closely, and refrigerate, for a few days

PRIME SEASON: (Indicated by darker shade).

| JAN | FEB | MAR | APR | MAY | JUN | JUL | AUG | SEP | OCT | NOV | DEC |

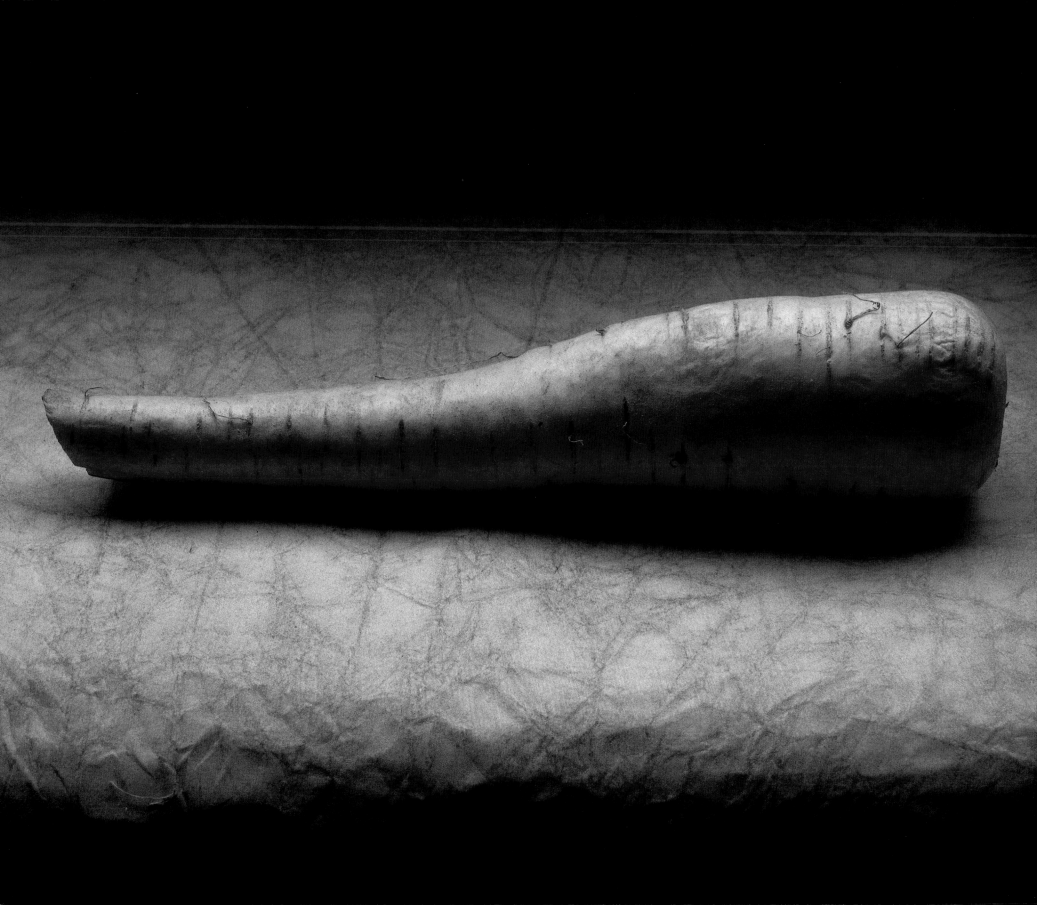

PARSNIPS

Parsnips normally taper to very slender tips that are often cut or broken. This is not a quality problem.

If they look like carrots, it's because family resemblance will out (sometimes, at least). If people could be induced to practice a little more adventure, parsnips might become almost as popular as their carrot cousins. Indeed, parsnips were more popular than carrots throughout the Middle Ages, but their status steadily declined until the eighteenth century, when, curiously enough, the American potato edged them out of the running. Nowadays the British and Eastern Europeans eat them once in a while, French and American cooks will throw them into a soup or *pot-au-feu*, Italians are said to fatten the hogs for their famous hams with parsnips, and that's about it. But at least they are available, for those who crave them and for those who like to experiment. Parsnips are wonderfully sweet and richly flavored; they deserve more attention. For those who have heard that parsnips are poisonous there is good news in two forms. First, the really ugly stories come from confusing wild parsnips with water hemlock. Second, while parsnips probably are slightly toxic, so are a remarkable number of other common foods.

Parsnips are cold-weather vegetables. They don't even taste right until they have spent some time at near-freezing temperatures, which convert starch to sugar. Modern techniques make them available all year, but parsnips are neither plentiful nor at their best in summer. Judge parsnips as you would carrots–they should be plump and fairly crisp. Avoid cracks, discoloration, dull, dry skin, shrivel, or softness. The medium sizes are tenderest and sweetest. Parsnips can be wrapped and refrigerated for a few days if necessary.

For those who cannot decide whether they like parsnips, vegetable soups and purées are the places to put them, where they will lend a subtle sweetness and nuttiness. For fans of parsnips, the roots can be steamed or simmered until tender, peeled, and finished with butter or a brown sugar or fruit glaze. Parsnips may be sautéed, creamed, French-fried, and even made into fritters or croquettes. They are most often served with roasts of any sort; they can be enhanced by sweet spice, fresh herbs, or orange or lemon flavor.

WHAT TO LOOK FOR:
plump, crisp

WHAT TO AVOID:
cracks, discoloration, dull, dry skin, shrivel, softness

HOW TO STORE:
wrap and refrigerate for a few days

PRIME SEASON: (Indicated by darker shade).

| JAN | FEB | MAR | APR | MAY | JUN | JUL | AUG | SEP | OCT | NOV | DEC |

PASSION FRUIT

These passion fruits are a little 'green' and need a few days of ripening.

First, the good news–passion fruit is good to eat. It has a rich, exotic, tropical flavor that makes for a pleasant change of pace during its short seasons. The bad news is for everyone who suspected that passion fruit is sexy. In truth, it gets its name from its flower, which is supposed to display the iconography of the passion of Christ, specifically the crucifixion. The devout can identify the three nails, the sponge soaked in vinegar, the wounds, the crown of thorns, and the apostles. But the hedonists among us need not be disappointed, for passion fruit is sensuous enough for any taste.

Passion fruit is shipped to temperate zones in very limited quantities from a variety of tropical sources; therefore seasonal availability is difficult to predict. The best chances are in midsummer, mid-fall, and late winter. Look for fruit that is a deep, dusty purple color. When properly ripened it will be a bit uneven and lumpy. It should yield to gentle pressure but still be firm overall. Avoid fruit that is pale in color or soft and bruised. Fruit that is underripe will improve some from aging at room temperature in a paper bag with a few holes in it. Once ready, it should be used immediately or refrigerated for a day or two.

The edible part of passion fruit is the greenish yellow mass of pulp and seeds within the shell-like skin. The fruit can be halved and eaten with a spoon just as it is or with a splash of lime juice. Imaginative cooks can find interesting ways to use passion fruit to garnish a variety of desserts, especially fruit salads. When the pulp is sieved and the seeds discarded, the juice of passion fruit lends a tropical touch to fruit punch (even the commercial one) and other beverages. It can certainly be used for ices and ice creams, mousses, bombes, and creams, soufflés, and sauces. Passion fruit can also be made into jelly and other 'put-up' foods.

WHAT TO LOOK FOR:
dark purple, lumpy skin yielding to pressure but still firm

WHAT TO AVOID:
pale and rock-hard or mushy, bruised fruit

HOW TO STORE:
at room temperature, then refrigerate for a day or two if necessary

PRIME SEASON: (Indicated by darker shade).

| JAN | FEB | MAR | APR | MAY | JUN | JUL | AUG | SEP | OCT | NOV | DEC |

PEACHES AND NECTARINES

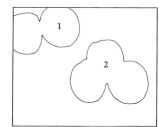

1. Peaches
2. Nectarines

'What a nectarine!' and 'nectariney-keen' never made it into English slang, not even 'nectarines 'n' cream,' though that fruit is just as fine and luscious as its well-honored sibling. No less an authority than Luther Burbank believed that peaches are a variant of the parent nectarine. A common notion is that nectarines are the product of a peach/plum cross. Not true. Peaches and nectarines are both ancient fruits from China that are so closely related that seeds from either of them can produce trees that bear either fruit (and occasionally both, or even–very rarely–both in the same fruit). In general, nectarines have a tendency to be smaller and sweeter than peaches, with a slight flavor difference that has been called vinous. The differences are growing slighter as modern agriculturists are breeding peaches with less fuzz. If peach fuzz disappears entirely, it will only be a happy memory of more heterogeneous times when variety was allowed to exist for its own sake.

Peaches and nectarines are summer fruits, at their peak in July and August. The early ones that start arriving in May are usually not as good, but the late fruits on the market into October should be fine. Winter imports from the Southern Hemisphere are scarce, expensive, and usually suffering from immaturity. Genuine tree-ripened peaches and nectarines are far too fragile to travel, so the practical ideal for harvesting is the mature-hard stage. Here the sugars are sufficient for eating later when the fruit has softened. Skin color is an important quality test. Avoid all traces of green. The ground color in white-fleshed varieties should be creamy, and yellow in the more common ones with orange flesh. The quantity and shade of blush are

results of the variety of the fruit and are not quality factors. Buy fruit that is firm and unblemished but not rock-hard. Sweet peaches and nectarines will be fragrant. In addition to immature fruit, avoid those that are mushy, bruised, or shriveled. Age fruit at room temperature in a paper bag with a few holes in it for one to several days. Sugars do not develop, but the texture and aroma will improve if the fruit is not too immature. Once softened and juicy, peaches and nectarines should be used promptly, but they can be refrigerated for a day or two if necessary. Both peaches and nectarines may be freestone or clingstone.

Peaches and nectarines are excellent breakfast, luncheon, snack, or dessert fruits eaten out of hand, cut, or cooked. They may be peeled or not, according to preference, though peaches frequently are. They are interchangeable for use in salads, pies, tarts, and other cooked desserts, or in ice cream, as garnishes for meats (especially duck and pork), and in relishes, preserves, and pickles. Peaches and nectarines may also be juiced or, more accurately, nectared.

WHAT TO LOOK FOR:
plump, large, firm but not hard fruit, creamy to yellow skin with varying degrees of blush, fresh fragrance

WHAT TO AVOID:
all traces of green, mushiness, bruises, shrivel

HOW TO STORE:
age at room temperature in a paper bag with a few holes in it until softened and sweet-smelling; refrigerate for a day or two if necessary

PRIME SEASON: (Indicated by darker shade). This may vary slightly due to local crops.

| JAN | FEB | MAR | APR | MAY | JUN | JUL | AUG | SEP | OCT | NOV | DEC |

PEARS

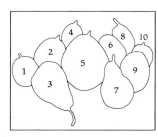

1. Seckel – popular with cooks, Seckels are also good eaten raw.

2. Yali

3. Bosc – has dull russet skin unlike the others. There may be yellow-rose areas as well.

4. Anjou – greenish with perhaps a bit of blush, becoming more golden when ripe. The gift-basket pear.

5. Chinese

6. Bartlett – the most popular in America, also highly prized in Europe where it is called Williams. Bartletts can be green with a trace of blush, or they can be of the newer red variety. The green ones become decidedly golden as they ripen.

7. Red Bartlett (Williams)

8. Comice – generally considered to be the finest of all pears on the market and usually the most expensive.

9. Red Comice

10. Fiorelli

The distinctive flavor of juicy ripe pears has made them popular for centuries, suitable even for special gifts to discriminating monarchs. And 'The Twelve Days of Christmas' has joyfully established the partridge in a pear tree' as one of the most comforting symbols of the winter holiday. Though partridges are tricky, or just expensive, to find, pears are available to all and relatively affordable.

The harvesting season is mostly fall, early and late, and some varieties keep well into winter. Thanks to cold storage there are pears on the market in spring and early summer, but you will probably not find one worth eating until the first good Bartletts arrive in August.

Avoid bruises, rot, moisture, and odd shapes. Pears ripen from the inside out, and fruits rot very quickly once they have reached melting softness. It is therefore best to buy firm but not rock-hard fruit of large size, characteristic shape and color, and with a minimum of blemishes. At this point pears are perfect for cooking. For eating, they should be allowed to mature at cool room temperatures until they yield to gentle pressure. To hasten this process, they can be placed in a brown paper bag with a few holes in it. Once ripe, use right away or store for a few days, refrigerated.

In addition to their virtues as fruits eaten out of hand, pears are successful baked, stewed, poached, pickled, in salads, relishes, and many dessert preparations. A wide variety of flavors and textures marry well, including sweet spices, vanilla, curry spices, chiles, nearly any other fruit, chocolate, sweetened wine with citrus and spice (for poaching), liqueurs, vegetables (especially Belgian endive), nuts, and many dairy products, notably ice cream

and almost any cheese. Good ripe pears are especially delicious with rich or ripened cheeses – Brie, Camembert, bleu – and with goat cheese.

WHAT TO LOOK FOR: large, firm, unblemished fruit, appropriate color

WHAT TO AVOID: bruises, rot, moisture, odd shapes, tiny rock-hard fruits

HOW TO STORE: ripen at room temperature, then refrigerate for a few days if necessary

PRIME SEASON: (Indicated by darker shade).

| JAN | FEB | MAR | APR | MAY | JUN | JUL | AUG | SEP | OCT | NOV | DEC |

PEAS

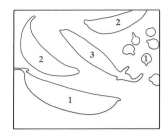

1. Garden (English)
2. Sugar snap
3. Snow

Peas have a wonderful gentleness about them, a softness that reminds the sentimental soul of spring rain. 'Sweet garden peas,' the bad restaurants say, and if we didn't know better we might be lured into visions of a dear little pea patch far from the hustle and bustle of real life. The rude truth is that if you want great peas you will either have to grow them yourself or buy them from someone else who grows them and picks them while you wait. This fantasy garden must be no more than five minutes from home, and the pot must be on and the diners must be at the ready. Peas are like sweet corn–their sugars begin to convert to starch immediately after picking, and they cannot taste perfectly fresh unless they are minutes or maybe a few hours from the vine.

Garden, or English peas are the most popular, the ones that are eaten so widely, especially in the U.S., Europe, and Central Asia. Second in popularity is the aptly named *mange-tout* ('eat it all') or sugar pea, which is also the Chinese snow pea, though it may have been developed in Holland. A fairly new arrival on the produce scene is a combination of the two, called sugar snap. These new peas have the best of both parents–tender/crisp

pods and sweet little peas too. Spring is the best time to find fresh peas, though snow peas are available daily, in shortest supply in midwinter. Choose peas that have small, bright green, glossy pods and fresh stem and blossom ends. Avoid dull, faded, or yellowish limp pods. Overgrown pods make for mealy garden peas and toughness in the others. Peas should always be used promptly, though they can be bagged and refrigerated for a day if necessary. The snow peas and sugar snaps on the market can be quite good if you buy them from a dealer who has fast turnover. Garden peas are usually overgrown and stale, hardly worth the hefty price. Test a pod by opening it with your thumb to see if the contents are green, small, and fairly even in size. Then decide if you wish to proceed.

Fresh garden peas are at their best when they are quickly steamed or simmered in a small amount of liquid and served with butter and perhaps a touch of fresh mint. Some people prefer them raw, right out of the pod. The French like to add a little shredded lettuce and onion and sometimes even carrot. A pinch of sugar will improve peas that are not perfect. Garden peas are popular vegetables for soups, salads, purées, stews, and vegetable combinations. In fact, they are the most popular vegetable in the West. In Chinese cooking, snow peas are quickly stir-fried to preserve their crunch, but these *mange-tout* can also be cooked in any ways that are suitable for garden peas. Simply trim the ends and string the sides if necessary. Sugar snaps can be used in any of these ways.

WHAT TO LOOK FOR:
small, bright green, glossy pods; fresh ends

WHAT TO AVOID:
dull, faded, yellowish, limp, overgrown pods

HOW TO STORE:
use promptly or refrigerate one day

PRIME SEASON: (Indicated by darker shade).

	JAN	FEB	MAR	APR	MAY	JUN	JUL	AUG	SEP	OCT	NOV	DEC
Snow peas												
Green peas and Sugar snaps												

CHILE PEPPERS

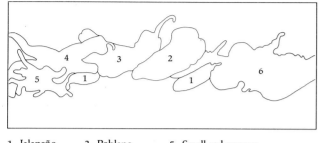

1. Jalapeño
2. Anaheim
3. Poblano
4. Hot pepper
5. Small red pepper
6. Serrano

If black pepper is the most ubiquitous seasoning from a vegetable source in the world, chile peppers are well in the running, not just as an assertive feature in the famous fiery cuisines of Mexico, India, and China, but also as a touch of interest to round out some of the most delicate dishes of man's creation. Before the discovery of the New World, only black pepper was available to add a little pain to the pleasure of dining. The new discoveries were fierier than anything known before, and they were latched on to with a vengeance.

There is medical evidence to indicate that chile habits in various climates are beneficial. In general, the hotter the weather, the spicier the food, causing perspiration, which keeps the system from knuckling under to tropical heat. In colder climates, a pinch of powdered cayenne might be as adventurous as the average cook gets, but even that timid gesture can wake up the flavor of *sauce hollandaise* or the elegant soup of mussel broth enriched with egg yolks and cream called *billi bi.*

Only in the areas of the world where chiles are an important part of cooking every day – Latin America, the southwestern U.S., southern Asia, Africa, and, to a smaller extent, the Mediterranean – will you find a good selection of fresh ones. Elsewhere, markets that cater to these tastes should have some selection of fresh chiles, but more dried varieties. Chiles are at their most plentiful from midsummer through fall, but some varieties are available most of the time. Look for plump, crisp, glossy, unblemished, fresh chiles. Avoid wilt, soft spots, or shriveled skin. Many fresh chiles are green, though all will ripen to red if left on the plant. Green chiles have a much longer shelf life than ripe ones, so green is preferred by grocers. Chiles in good condition can be refrigerated for up to two weeks. Green chiles tend to taste hotter than red, though they just lack the heat-balancing sweetness that develops with ripeness.

Buying fresh chiles can be confusing for two reasons. Not only do names vary, but the peppers themselves may vary in their flavor and heat because of growing conditions and cross-pollination. Produce vendors should be able to tell you whether or not the chiles are hot, but then you will have to test them for intensity and use the desired quantity. For safety's sake, taste the dish rather than the raw chile. Caps and membranes are always discarded, but the seeds, which are very hot in hot chiles, are sometimes used in cooking, though more often they are discarded. Soaking the flesh in vinegared water will draw out some of the heat. Some people are so sensitive to the volatile oils in hot chiles – the active ingredient is called capsicin or capsicine – that they must wear rubber gloves, even a mask, while cleaning them. Most people need only remember to scrub their hands and utensils well with soap and hot water before touching other foods or, especially, the eyes or other sensitive parts of the body. A painful burning sensation is the result of negligence. This happens only once to wise cooks.

Chiles are combined with a variety of foods, especially meats, poultry, seafood, eggs, cheese, tomatoes, onions, garlic, eggplant, squash, beans, corn, and sweet peppers or other chiles. They are cooked in several ways, especially sautéed, stir-fried, stewed, or roasted like sweet peppers. Chiles are also used raw and pickled. Tolerance to chiles is acquired. Spicy dishes that are relished by the

PRIME SEASON: (Indicated by darker shade). This may vary slightly due to local crops.

| JAN | FEB | MAR | APR | MAY | JUN | JUL | AUG | SEP | OCT | NOV | DEC |

accustomed can actually raise blisters on the mouth and lips of the uninitiated. Ice water will not put out the fire, but beer and very hot tea are rumored to douse the flames.

HOT PEPPER
This is the name in general markets, and it can include a variety of green, yellow, orange, and red chiles. The smallest ones may be very hot, while the larger ones are often milder.

JALAPEÑOS
They have a distinctive shortened shape and are usually marketed by name. They are hot. A very pale yellowish green chile of the same shape is popular in California.

POBLANOS AND ANAHEIMS
These are large, usually mild green chiles that are suitable for stuffing. One of them is customarily used for Mexican or Tex-Mex *chiles rellenos.*

SERRANOS
These are smaller green chiles, longer and slimmer than jalapeños and widely used in cooking. Serranos are hot.

Among the other chiles on the market are cherry peppers, rounded chiles of green, red, yellow, or orange that are often pickled and that vary in their heat; tiny, hot little tabasco peppers, the ones that are used for the condiment; and cayenne peppers, which are any of three different varieties of small red chiles, all of which, of course, are hot.

WHAT TO LOOK FOR:
plump, crisp, unblemished chiles

WHAT TO AVOID:
wilt, soft spots, shriveled skin

HOW TO STORE:
wrap loosely and refrigerate for up to two weeks, less for red chiles

SWEET PEPPERS

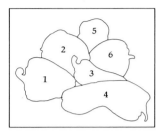

1. Yellow (Holland)
2. Green
3. White (Holland)
4. 'Italian' or 'frying'
5. Red
6. Purple (Holland)

One of the New World's major gifts to the Old is the pepper, probably misnamed by Columbus himself for black pepper, a familiar spice in Renaissance Europe. The two have nothing in common but zesty flavor. American peppers, capsicums, are part of the famous (sometimes infamous) nightshade family along with tomatoes, potatoes, eggplant, tobacco, belladonna, and deadly nightshade. Sweet peppers have rarely received the bad press suffered by their siblings–hot peppers are the ones accused of being lethal–but rather have been freely embraced throughout the world. They come in a lovely array of colors from near-white to pale green to very green, and from yellow to orange to red to purple. All sweet peppers are similar in flavor and texture–crisp and refreshing raw, pleasantly assertive when cooked to tenderness. They vary in their subtle balances of sweetness and delicate peppery quality.

The familiar green bell pepper is available year round, a bit more plentiful in summer. When ripened on the vine it turns red and gains sweetness, but it also costs more to produce and has a poor shelf life, reasons why the red version is costlier and rarer. The paler green, tapering ones called frying or Italian peppers have a bit more flavor than the bell-shaped ones and are in fairly good supply. The other peppers are comparatively rare, most of them grown in Italy or Holland, available in summer and through the fall, and are quite expensive but worth it. Many people find them very sweet and more digestible than their relatives. Choose peppers that are firm and glossy. Avoid bruises, rotten spots, and shrivel. Fresh peppers can be wrapped and refrigerated for a few days if necessary.

Sweet peppers are important vegetables in many parts of the world, especially in Italy, Spain, Eastern Europe, and the American South. They are freely combined with all manner of raw ingredients, especially beef, veal, chicken, sausages, anchovies, onions, garlic, eggplant, squash, tomatoes, beans, rice, corn, and olive oil. They can be stewed, sautéed, fried, baked, stuffed, pickled, dried to make paprika, or eaten raw, particularly in salads. One of the finest flavors is achieved by roasting peppers over a hot flame, grill, or under a broiler until the skin is uniformly charred. Many cooks then cool them in a paper bag, thus steaming the skin to further loosen it. After the peppers are peeled and cleaned, their flavor is wonderfully rich and smoky.

WHAT TO LOOK FOR: firm, plump; glossy skin

WHAT TO AVOID: bruises, rotten spots, shrivel

HOW TO STORE: refrigerate for a few days if necessary

PRIME SEASON: (Indicated by darker shade).

| JAN | FEB | MAR | APR | MAY | JUN | JUL | AUG | SEP | OCT | NOV | DEC |

PERSIMMONS

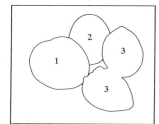

1. Fuyu
2. Italian
3. Hachiya

For residents of the southern U.S. and thereabouts, in areas not too urban, one of the great joys of autumn is the appearance of a rather strange wild fruit, the native American persimmon. It is a small thing, hardly larger than a walnut, that grows on large trees and has an extraordinary tropical sweetness. But timing is crucial, for fruit eaten only the merest hint underripe starts out sweet but then develops a powerful astringency that threatens to turn the mouth inside out. This goes for the large commercial persimmons, too, all of which were bred from Japanese forebears. These Japanese fruits (which probably originated in China) are the ones known and enjoyed around the world, especially in Italy and southern France, as persimmons or kaki.

Persimmons begin to arrive in late September and are usually available into January. Their shape can be tomatolike or rather flat or heart shaped.

Whatever the size or shape, there is always the distinctive four-leafed cap and an orange color that ranges from yellowish to reddish. Some have seeds while others do not. Table-ripe persimmons are very fragile, so buy unblemished fruit that is slightly soft to the touch and ripen it at home. To hasten removal of the dreaded astringency, you might want to try a well-sealed bag without the usual holes. Ripening time can vary from a few days to a week or more. Some persimmon varieties are ready to eat when they reach a pleasant softness, but others require continued ripening until the skin begins to shrivel and the flesh becomes jellylike, translucent. Tasting is the only test, so never take the chance of spoiling a meal or scaring family or guests away from this delectable treat.

The native American persimmon is most often used for the Southern favorite, persimmon pudding, a baked thing like a very moist cake with a unique and pleasant waxy texture. The commercial variety will work here too, sans texture, as well as for stewing, jams, jellies, and preserves, cakes, pies, ices, and ice creams. This so-called Japanese persimmon is more often eaten out of hand or with a spoon or knife and fork (the skin is edible, but usually discarded). It can be included in salads, either mixed-fruit concoctions or artfully composed presentations of greens, poultry, and such. Persimmons can go almost anywhere a mango might be welcome.

WHAT TO LOOK FOR:
plump, evenly orange fruit, either shiny or waxy-looking

WHAT TO AVOID:
cracks, splits, moisture, greenness, rock-hard fruit

HOW TO STORE:
ripen in a bag at room temperature, then refrigerate for a few days

| JAN | FEB | MAR | APR | MAY | JUN | JUL | AUG | SEP | OCT | NOV | DEC |

PINEAPPLES

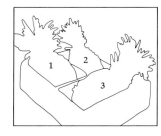

1. Hawaiian
2. Mexican
3. Honduran

This Hawaiian pineapple is in excellent condition. The others are typical, fresh enough, and would need sweetening to bring out their flavor.

Of all the New World discoveries of Columbus and his crews, pineapples were the fruits that caused the biggest stir back home, the ones that received the strongest efforts at cultivation under alien conditions. With characteristic greenhousing skill, one Dutch gardener in Leyden actually succeeded, but the results were neither economical nor superior enough to warrant a lasting industry. However, conditions in Hawaii proved to be magical and led to the largest pineapple industry in the world. Pineapples may well be the most-loved tropical fruit.

Pineapples are grown in many other tropical areas, especially Puerto Rico, the Philippines, and neighboring parts of Southeast Asia. Only a small percentage of the crop is shipped fresh (usually by air) to distant markets because of perishability and ripening problems. Pineapples can only ripen on the plant, having no starch reserves to convert into sugar after they are picked. Ripe pineapples are very fragile and, with a shelf life of only a few days, cannot travel. So pineapples to be shipped must be picked while they are immature, and the degree of immaturity is the crucial quality factor. Hawaiian pineapples are the largest and usually the best ones, with rind color that shows yellow and golden-brown mixed with the remaining immature green. Others are likely to be shorter, fatter, and greener. As one would expect, the leaf crown should be fresh and green, and leaf pulling is not a way to judge ripeness. More-mature pineapples normally have a crown that is shorter in proportion to the fruit than green ones'. The most

reliable quality test is fragrance. A pineapple that smells fresh and sweet will also taste that way. One with no odor will have little taste. Once picked they cannot improve, though some people like to age pineapples for a few days, during which time the acid decreases and gives the illusion of sweetness. This practice is not recommended. In addition to dry, browning crowns, avoid soft spots, moisture leakage from the bottom, and a fermented odor. Fresh pineapples should be used promptly, though they can be stored for a day or two if necessary, refrigerated or not.

Pineapples can be cut in several ways to yield rings, chunks, sticks, or boats. There are many possibilities for salads, desserts, beverages, main courses, and preserves and relishes. For meat dishes, pineapple seems to go especially well with pork, either in a Polynesian stir-fry or the familiar American baked ham. The core of a ripe pineapple is edible, and in greener ones it makes a refreshing munch even if the fibers are too tough to swallow. The bottom part of the pineapple will be noticeably sweeter than the top. Cooks should know that raw pineapple contains an enzyme that interferes with gelatin's ability to gel. A small amount of cooking eliminates this effect.

WHAT TO LOOK FOR:
large fruits with fresh crowns, a minimum of green to the skin, and a fresh, sweet smell

WHAT TO AVOID:
dry, browning crowns, very green skin, softness or fermented odor

HOW TO STORE:
use promptly or hold for a day or two, refrigerated or not

PRIME SEASON: (Indicated by darker shade).

| JAN | FEB | MAR | APR | MAY | JUN | JUL | AUG | SEP | OCT | NOV | DEC |

PLUMS

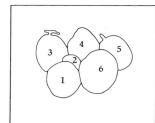

1. Empress
2. Roysum
3. Italian prune
4. Kelsey
5. Angelino
6. President

What is native to Europe, China, and North America, freely interbred, grown in all temperate zones (every continent but Antarctica), and famous for its juicy sweetness? The plum, the most diverse of the stone fruits. It is no accident that political appointments and other sweet situations are named after this cousin of peaches and nectarines. Plums come in a wide range of sizes, shapes, and colors. They can be as small as a cherry or as large as a baseball, either very round, elongated, or heart shaped, and sometimes deeply indented on one side so that they recall, to minds so inclined, buttocks. Color varies from yellow to green to red to blue to purple to black with combinations in between. The flavor of a good ripe plum has a luscious sweet/tart balance that distinguishes it as one of the finest fruits.

More than a dozen plum varieties regularly make it to market, starting in May for the early varieties, peaking in July and August, and finishing up with the late varieties into October. In addition to the well-known varieties, the Santa Rosa and the Italian prune, there are wonderfully charming names including El Dorado, Friar, Elephant Heart, Queen Anne, Washington, Jefferson, and Reine Claude (Green Gage). Some South American imports are on the market in cold weather, but they tend to be expensive and sourish from early picking. In fact, most plums are picked too green so that, as usual, local tree-ripened crops are best. Green plums should be yellowish and the others should be deeply colored. Underripe fruit should never be refrigerated but allowed to mature at room temperature.

In addition to eating out of hand or in fruit salads, plum fanciers find many ways to enjoy the fresh fruit in pies and tarts, cakes, puddings, ices and ice creams, mousses, bombes and creams, and stewed or poached preparations. There are also many possibilities for preserves, jams and jellies, relishes, even spirits, which are often made with varieties too tart to eat fresh–Damsons, Sloes, and Beach plums. Fresh or preserved, plums go especially well with sweet spices, cream, nuts, citrus and other fruits, pork, duck, and goose. *Prune* is a term applied to some fresh plum varieties, but it most often means a dried plum, whose uses go far beyond geriatric breakfast ritual to some of the glories of Eastern European and Scandinavian cookery.

WHAT TO LOOK FOR:
soft, plump, smooth skinned; deep, even color with waxy bloom intact

WHAT TO AVOID:
rock-hard fruit or shriveled skin, bruises, rot, cracks

HOW TO STORE:
refrigerate very ripe fruit for use in a few days

PRIME SEASON: (Indicated by darker shade).

| JAN | FEB | MAR | APR | MAY | JUN | JUL | AUG | SEP | OCT | NOV | DEC |

POMEGRANATES

One of the sexiest of fruits, the pomegranate has enjoyed an honored place in fertility rites and on the tables of those who take seriously the sensuality of eating. All the peoples of the Middle East and the Mediterranean have celebrated its properties, culinary and otherwise, since prehistory. And at least one modern sybarite has long fantasized the perfect orgy–a truckload of pomegranates, a roomful of compatible people, and time to see what develops. Only the insensitive could fail to respond to the beauty of the external form, the luxuriant fertility of the interior, and the rich sweet flavor of the juice. It has a berrylike flavor (it *is* actually a berry) that is at once unique and exactly the way a red fruit should taste. Grenadine syrup was originally made from pomegranate juice, but no one is telling what it is made from now.

Pomegranates come to market in the fall, peaking in October and November with some availability in late September and December. Choose large, bright-skinned fruit that feels heavy for its size. Most of the fruits on the market are decidedly red skinned, although there are paler, more golden varieties, even green or white ones. Avoid a dull shriveled look or any rot or breaks in the leathery rind. The six-pointed blossom end is very handsome, also fragile. For decorative purposes, try to find fruit that has not been knocked about. Fresh pomegranates keep well for at least a week or two, refrigerated, and the late ones can even be held into January.

Most often pomegranates are eaten out of hand with care to avoid the bitter white pith and squirting the juice onto clothing. It stains. Many people feel that the best way to avoid this problem is to avoid clothing, rather than pomegranates. Some like to munch on the actual seeds, while others discard them. The juicy little red pellets can be sprinkled over salads, cold meats, and desserts to add flavor and texture. There is at least one traditional Lebanese hors d'oeuvre, the sesame- and garlic-flavored eggplant paste called *baba ghanooj*, that is brightened by a pomegranate garnish. The fruit can also be juiced like a citrus fruit for ices, jellies, and sauces, and for marinades and glazes for poultry.

PRIME SEASON: (Indicated by darker shade).

| JAN | FEB | MAR | APR | MAY | JUN | JUL | AUG | SEP | OCT | NOV | DEC |

POTATOES

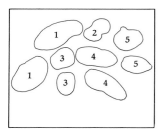

1. Russet (Idaho)
2. Red
3. New round white
4. Mature long white (Irish)
5. Mature round white

The largest food crop in the world, by most reckonings, was also among the slowest of the now-popular discoveries in the Americas to gain acceptance elsewhere. Early Spanish explorers found potatoes in their native Peru and Bolivia and probably brought them back to Europe then, but only in the eighteenth century did potatoes finally begin to find a European or U.S. audience. The most famous name in French potatoes, Antoine Parmentier, is credited with the clever ruse of providing guards for his new potato field during daylight hours only, encouraging theft at night. This did much to popularize potatoes among the folk, and he also presented a bouquet of potato flowers to Louis XVI, who is said to have worn one on his lapel. Potatoes had arrived in France. Ironically, the great famine relief for Ireland became the single staple, so that when a fungus attack blighted the potato crop in 1845 and '46, an estimated two and one-half million people died, while countless others emigrated. Such is the role of agriculture in history.

Potatoes are plentiful all year. Starting in the spring and lasting through summer, there is a larger availability of 'new' or early potatoes, which are actually harvested while immature. These should have very thin, fragile, glossy skin that might be rubbed in spots. Mature potatoes are duller and dusty-looking. All potatoes should be plump and firm and of fairly regular shape with no cuts, bruises, shrivel, or sprouts. They can be stored for a week or two at room temperature, or even longer if there is a cool, dry, dark, well-ventilated place for them. Refrigeration is undesirable because it converts starch to sugar, and light causes greening, an alkaloid called solanin that is toxic in quantity. The green may be simply peeled away, but the remaining potato will not be at its best. There

is some indication that potatoes should be stored separately from onions to avoid detrimental gas exchanges. Should potatoes begin to sprout, they may still be used, so long as the potato is still firm and the sprouts are removed. Potatoes with internal discoloration are either blighted or damaged in storage, and should be complained about.

Mature potatoes, especially russets, called Idahoes in the U.S., are relatively dry and starchy and give the best results for baking, frying, some casseroles, and mashing. Potatoes should always be mashed or riced by hand, because purées achieved with a mixing machine give gluey results. New potatoes are moist and waxy, therefore best for boiling, steaming, and salads, though some people fry them too. The perfect salad potato hardly exists anymore, but the freshest 'news' will give the best results. All potatoes must be cooked or placed in water immediately after peeling to avoid discoloration. Brown-, beige-, and red-skinned potatoes with white or creamy flesh are most common, but there are also potatoes with blue, black, or pink flesh and, most notably, the highly prized Finnish or Dutch yellow potatoes.

WHAT TO LOOK FOR:
plump, firm, regularly shaped potatoes; fairly clean, thin, fragile, glossy skin on new potatoes

WHAT TO AVOID:
cuts, bruises, shrivel, sprouts

HOW TO STORE:
eliminate any blighted potatoes and store, with air circulation, in darkness, for a week or two at room temperature or longer in a cool, dry, dark, well-ventilated place

PRIME SEASON: (Indicated by darker shade).

| JAN | FEB | MAR | APR | MAY | JUN | JUL | AUG | SEP | OCT | NOV | DEC |

PRICKLY PEARS

The cactus is one of the desert plants that needs no oasis to sustain it; in fact, it is something of an oasis in itself, providing watery flesh that quenches thirst, and another thirst quencher too–tequila. While the green pads of some varieties–Nopales and Opuntias–are used as a vegetable, the fruit of these distinctive spiny plants is gaining a wider following. Still a luxury item in most parts, these rather barrel-shaped oddities in greenish or reddish hues, called cactus pears or prickly pears (for obvious reasons), can sometimes be found in fancy markets. They have a fresh, fruity taste and a refreshing quality not unlike the plant itself. Mercifully, the thorns are removed so that shoppers need not suffer the pain inflicted by these tiny barbs that can work their stubborn way into the fingers of the unwary and the tongues of greedy children.

Prickly pear season starts in fall with some availability through midwinter. Choose fruit in good condition without wilt, cracks, or rot. It should be fairly firm but not hard. When ready to eat, green varieties become yellowish and red varieties lose some green and become more uniform in color. They also soften a bit. There is no substitute for ripening on the plant, but prickly pears are picked green for transportability and will benefit to some degree from aging at room temperature until they reach the desired state. They may then be stored, refrigerated, for a few days.

Prickly pears are always peeled, the skin discarded. The flesh, with its seeds, is eaten in salads or as is. A spritz of lime juice is a welcome touch. There are many salad combinations to be explored–greens, poultry, shellfish, nuts, and so on. And there are also possibilities for jams, jellies, pickles, and preserves, as well as ices and mixed fruit desserts.

WHAT TO LOOK FOR:
firm but not hard, reddish or greenish

WHAT TO AVOID:
withering, cracks, rot

HOW TO STORE:
mature at room temperature, then wrap and refrigerate, a few days

PRIME SEASON: (Indicated by darker shade).

| JAN | FEB | MAR | APR | MAY | JUN | JUL | AUG | SEP | OCT | NOV | DEC |

QUINCE

These quinces are at an ideal purchasing stage of firmness and freedom from blemishes.

The quince has just as noble and ancient a history as its relatives, apples and pears, but it has sadly lost popularity to the point where most people think of it as merely an old-fashioned curiosity, if they think of it at all. A victim of fashions in food, the quince has slipped from among the most useful of fruits (in medieval Europe) to among the most neglected. It is all a matter of sweetness–moderns seem to have lost appreciation for tart fruits, rather expecting them all to be sugary. In fact, there are quince varieties in the world that are quite pleasant and refreshing just as they are, just as they were enjoyed for centuries, while the more usual ones require a minimum of cooking and sweetening to bring out their agreeable flavor, which is distinctive but more like apple and pear than anything else.

Quinces are available in the fall. The larger ones are usually best. Color ranges from green, unripe, to yellow, when the fruit ripens. Very ripe quinces are quite fragile, so a small amount of scarring is to be expected. It is therefore best to choose fruit that is firm but not hard, with skin on its way to yellow but still in excellent condition and resistant to bruising. Quinces can then be ripened at home in a paper bag with a few holes in it until they are quite

yellow and beginning to soften. This method avoids the inevitable rough treatment in the market. Never buy badly bruised ripe fruit, for decay spreads rapidly. Ripe quinces may be stored in the refrigerator for a few days if necessary.

Quince is the original fruit for marmalade, which takes its name from the Portuguese word for quince, *marmelo*. There are many jam, jelly, and preserve possibilities for which the quince's high pectin and acidity levels are ideal. The fruit may also be baked, stewed, or sautéed with little or no sweetening added to create a tasty foil for fatty flesh–pork, goose, and duck. The delicate pale pink of cooked quince is appealing to the eye. Pies and tarts are also traditional, as is teaming the quince with sweet spices, apples, citrus, grapes, and berries of all sorts.

WHAT TO LOOK FOR:
large, firm, yellowish skin

WHAT TO AVOID:
overripe, badly bruised fruit

HOW TO STORE:
ripen in a paper bag at room temperature, then refrigerate for a few days if necessary

PRIME SEASON: (Indicated by darker shade).

| JAN | FEB | MAR | APR | MAY | JUN | JUL | AUG | SEP | OCT | NOV | DEC |

RADISHES

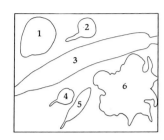

1. Chinese
2. Black
3. Daikon
4. Round white
5. Icicle
6. Round red

As garden vegetables go, radishes are plentiful enough, but they rank high on the list of versatile edibles that are woefully underused, especially in the U.S. Radishes are known chiefly as a salad ingredient or a 'decorative' thing carved into unimaginative or sloppy shapes that can guarantee generations of apathy. Europeans know better, but not much. Escoffier dismissed radishes as decorations for other hors d'oeuvres, or a separate thing garnished with butter curls, neglecting to mention that fresh radishes with crusty bread and fine sweet butter make one of the best appetizers on earth. More respect is paid in Asia, where radishes of various sorts are kitchen staples.

Radishes vary in their keeping qualities, so they are classified as summer or winter. Summer radishes are the small ones of red, white, or red and white. They may be round or elongated. These are the more perishable ones that are most plentiful in spring, though they can be found nearly any time. Look for small radishes of regular shape with smooth, glossy skin. Avoid cracks, sprouting, soft spots, and rot. Oversize summer radishes can be tough and woody or hollow and pithy, and strongly flavored. Radishes marketed in bunches with greens attached are an excellent, if expensive, choice, because fresh greens assure the buyer of fresh, fine-flavored radishes. Radishes can be refrigerated for a few days if necessary. Bunch radishes must have their tops removed before storage to avoid moisture and nutrient loss.

Winter radishes may be black or white or white shot with green or other colors. The black ones can be turnip shaped and sized or elongated, and the white ones can grow to be huge, a yard long or more, and either thin or fat. They all store well once the greens are removed. Radishes should be firm and unblemished.

Though summer radishes can be sliced or chopped and added to salads, many people feel that they are best enjoyed whole. Leaving an inch of the stems attached (bunch radishes, of course) makes a convenient and edible handle. In addition to bread and butter, salt is an ancient accompaniment. Summer radishes are much neglected as cooked vegetables. They are delicious steamed or braised in butter and served with roast meats. They can be used in any recipe for turnips.

Black radishes have a pungent flavor and should be used sparingly, raw, cooked, or marinated. Daikons, the long white Japanese ones, and Chinese radishes, the fat ones, are milder in flavor. They can be cooked and served in pieces or puréed. Raw Asian radishes are delicious when finely shredded for salad or as an accompaniment to sushi and sashimi. Daikon pickles are common munches to end a Japanese meal. Both simple and elaborate Daikon carvings grace Asian tables for special occasions. All radish leaves are edible, as potherbs or used sparingly in soups or salads.

WHAT TO LOOK FOR: firm, plump, unblemished radishes, glossy skin; summer radishes should be small; fresh leaves, if any

WHAT TO AVOID: wilt, cracks, sprouting, soft spots, rot; wilted leaves, if any

HOW TO STORE: remove leaves, refrigerate wrapped for a few days if necessary

PRIME SEASON: (Indicated by darker shade).

| JAN | FEB | MAR | APR | MAY | JUN | JUL | AUG | SEP | OCT | NOV | DEC |

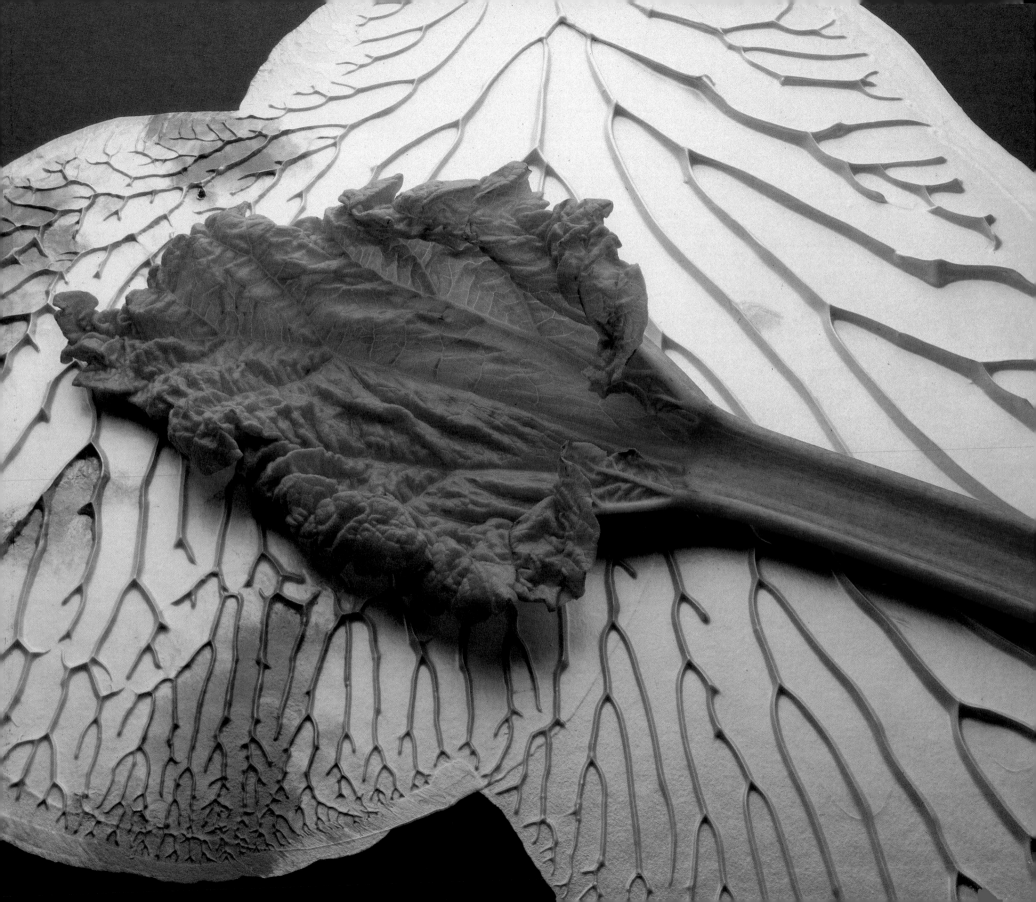

RHUBARB

'D'ya think the rain will hurt the rhubarb?' That line broke a lot of ice and broke up a few vaudeville audiences in its day. That day is past, but rhubarb is still with us, much of it hothouse-grown these days so that there are no rains to damage the leafstalks by spreading disease. Rhubarb got its name when the ancients observed that it was grown along the Rha river, now known as the Volga, by the resident barbarians. These Siberian Tartars had a reputation for incivility to their neighbors, which may explain how rhubarb came to be the name for a modern-day fracas on the baseball diamond. Despite its funny name and comical associations, rhubarb is a delightful vegetable usually used as a fruit, with an assertive tartness that perks up sweets. Not by accident is it also known as pieplant.

There is a little bit of rhubarb growing all the time, but January is the month when the first general availability begins. Expect it to peak in April and May and finish for the year in August. Field-grown rhubarb is large and dark red with dark green leaves. Deeper color usually means more flavor, but field-grown rhubarb may be stringy and require peeling. Choose crisp, plump, medium-sized stalks of red or pink with fresh-looking leaves, if any. Rhubarb stalks may be better than two feet long, or they may be cut into pieces. Avoid wilt, blemishes, and stalks that are skinny or overgrown. When very fresh, rhubarb should keep well for a few days if wrapped and refrigerated.

Cooks should be warned that rhubarb leaves are rich in oxalic acid and other poisons and must never be eaten. They are better suited to tanning hides and cleaning aluminum. The stalks should be tested for tenderness by snapping. If stringy they should be peeled. Rhubarb is most often sliced, well sugared, and stewed or baked into pies or tarts. Strawberries are traditional partners. Rhubarb can also be made into sauces, puddings, and preserves and relishes of various sorts. The juice can be used for beverages, even wine. In the Middle East rhubarb is sometimes used unsweetened to add tartness to meat or vegetable dishes, such as stuffed grape leaves and ground lamb preparations. There are reputed to be people in remote areas who eat rhubarb raw.

WHAT TO LOOK FOR:
crisp, plump, medium-sized stalks of red or pink, fresh-looking

WHAT TO AVOID:
wilt, blemishes, skinny or overgrown stalks

HOW TO STORE:
wrap and refrigerate for a few days if necessary

PRIME SEASON: (Indicated by darker shade).

| JAN | FEB | MAR | APR | MAY | JUN | JUL | AUG | SEP | OCT | NOV | DEC |

ROOTS AND TUBERS

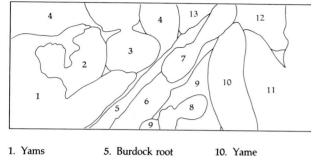

1. Yams
2. Apia
3. Jicama
4. Malanga
5. Burdock root
6. Salsify
7. Lotus root
8. Taro
9. Yautia
10. Yame
11. Cassava
12. Jerusalem artichoke
13. Water chestnut

Trial and error is the story of human nutrition. Whoever got the notion that the underground parts of plants might be good to eat set off a round of experiments that took us from the bland to the sweet to the pungent to the downright poisonous. An excellent example of man's ingenuity and persistence can be found in taro and cassava, where those who lived to tell about testing them raw discovered that cooking rendered them harmless and nutritious. Roots and root swellings of dozens of plants are staple crops throughout the world. Some of them provide life-sustaining starch, while others are enjoyed occasionally or sparingly for their crunch or savor.

Roots are harvested at various times of the year and tend to keep quite well, compared with leafy vegetables. They are often gnarled and warty, but they should always be firm, plump, and crisp. Avoid softness, shrivel, moisture, blemishes, mold, or sprouting. The term 'root cellar' is apt, for most roots keep best in a cool, dry, dark, well-ventilated place. They can also be kept at cool room temperature or refrigerated for use within the week. Most roots and tubers require air circulation and should not be tightly wrapped.

TRUE YAMS
They are of African origin and are usually very large, with moist flesh, and can be used for any purposes where sweet potatoes are good. Yams are rare in Europe and the U.S.

APIA
It is one of the starch roots popular in the Caribbean.

JICAMA
This is a fat root with a crisp texture like water chestnuts that is becoming popular as a cooked vegetable and as a raw salad ingredient.

MALANGA
It is an important plant in tropics of the New World and Africa, where its large tubers provide staple starch and the leaves, called callaloo, are also eaten, especially in a stew also called callaloo. Malanga root is something like a moist white sweet potato, very starchy, and it must be cooked before eating to dispel considerable poisons.

FRESH WATER CHESTNUTS
These are occasionally available, especially in Asian markets. They are peeled and sliced, blanched slightly, and then used in traditional Asian dishes or salads or with any other vegetables where their crunch would be interesting, even Brussels sprouts.

BURDOCK ROOT
This is much prized in Japanese cookery, where it is known as *gobo* and is often pickled. Americans and Europeans know burdock best as the pesty weed with the prickly hitchhiking burrs that are so hard to remove from clothing and dogs' fur.

LOTUS ROOT
This one is the curious thing in Asian markets that looks like a straight string of large brownish sausages. When trimmed and cut, it has a delicate flavor and crunch and an elegant appearance in Chinese cooking.

SALSIFY
This is the name for several roots, one of which is actually a black radish. A more lightly colored one is known as oyster plant, though few people sense

PRIME SEASON: (Indicated by darker shade).

| JAN | FEB | MAR | APR | MAY | JUN | JUL | AUG | SEP | OCT | NOV | DEC |

the taste similarity these days. Salsify can be buttered, sauced, or marinated and is a great delicacy to those who fancy it. It is losing popularity in Europe and was never very important in the U.S.

TARO

It is a popular tropical Asian starch root from which is made the famous Hawaiian concoction *poi*. Like malanga, yautia, and cassava, it must be thoroughly cooked. Yautia is similar to malanga and dasheen, which it is also called. They are used interchangeably.

YAME

This is a big, fat starchy tuber from the Caribbean that is used as a potato substitute.

CASSAVA

This is another important, and poisonous, starchy root, which when cooked is the source of tapioca, the ubiquitous Brazilian meal called *farinha de mandioca*, and a liquid derived form the milky juice called *cassareep*, which is used to flavor sauces and stews.

JERUSALEM ARTICHOKE

It is also called sunchoke. It is the tuber of an American member of the sunflower family. It has nothing to do with Jerusalem; the name is a corruption of the Italian *girasole*, which means 'sunflower.' The roots are crunchy and pleasantly bland with a slight sweetness. They may be used raw anywhere that water chestnuts would be good, or as a cooked vegetable, peeled or not.

WHAT TO LOOK FOR:
firm, plump, crisp

WHAT TO AVOID:
softness, shrivel, moisture, blemishes, mold, sprouting

HOW TO STORE:
cool, dry, dark, well-ventilated place (root cellar) or refrigerator, loosely wrapped, for up to a week

SCALLIONS

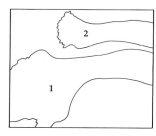

1. Bulbing (Italian) scallions
2. Standard green onions

'Spring onion' is one of their names, an engaging notion that is fairly accurate when we consider that scallions are pulled young, while they are still tiny and tender. Just exactly what a scallion is, though, creates some problems. 'Green onion' is probably a safer term, for the majority of the scallions on the market are simply immature white onions harvested before they have started to bulb or while the bulbs are still small. And yet a scallion might also be a young shallot or even another member of the onion family erroneously called Welsh onion. Whatever the name or botany, scallions have a fresh and peppery onion flavor that is delicious and versatile.

Scallion production is by no means limited to spring, though the supply is greatest from then into summer. Scallions are plentiful all year. Look for fresh blue-green leaves that are not trimmed excessively. The stems may be straight, most commonly, or bulbous. Avoid wilt, rot, yellowing, and watery bruising. Scallions are far more perishable than dry onions and can only be stored for a few days, wrapped and refrigerated.

In the West scallions are often eaten raw as *crudité* or in salads. Many people fail to realize that they are successful cooking onions too. In Chinese cooking they are often stir-fried or sliced or shredded and strewn on steamed or fried fish dishes. Japanese soups and sauces are often finished with a sprinkling of freshly sliced scallion greens. Asian cooks also cut scallions into decorative flowers and useful brushes for sauce dipping (especially for Peking duck service) that represent edible garnish at its simplest and best. In Europe there is a limited tradition of scallion cookery, mostly in custard or cheese tarts, or occasionally in soups. Bulbed, or Italian scallions, work well for these purposes. In the U.S., only in Louisiana is the scallion a cooking staple. This tradition seems to date back to the earliest French settlements, where shallots were unavailable and scallions were substituted. To this day, traditional Creole parlance labels scallions as shallots and confuses cooks from other parts of the country. American cookbook authors have often recommended the white part of scallions as a shallot substitute, and while it is no substitute, it is tasty. Scallions should always be sliced with a very sharp knife and a sawing motion to avoid bruising the tender leaves and stems.

WHAT TO LOOK FOR: fresh blue-green leaves that are not trimmed excessively

WHAT TO AVOID: wilt, rot, yellowing, and watery bruising

HOW TO STORE: wrap and refrigerate for a few days if necessary

PRIME SEASON: (Indicated by darker shade).

| JAN | FEB | MAR | APR | MAY | JUN | JUL | AUG | SEP | OCT | NOV | DEC |

SHALLOTS

Greening is generally considered to be undesirable, but it occasionally shows up with no ill effect in shallots that are in otherwise perfect condition.

Shallots have a well-earned reputation for being the most elegant of the 'kitchen lilies' (the onion family) largely because of their important place in classical and modern French cooking. Only recently have they gained a widespread following in the rest of the world. Their subtle complexity has often been described as a combination of onion and garlic, but the only accurate appraisal is that they have a distinctive flavor, the flavor of shallot.

Occasionally marketed in their fresh green scallionlike state, the majority of shallots are available, year round, as dry bulbs. They grow in multiple cloves, but unlike garlic, the group is not surrounded by exterior layers of skin. The most characteristic configuration is two separate cloves attached at the root. Shallots range in size from about a half-inch long for 'true,' or French shallots, to two inches long for newer varieties. They are interchangeable, though most experts find the small ones finer. Look for dry, papery skin of greyish to reddish brown with a light sheen to it. Cloves should be plump, firm, and heavy. Avoid softness,

moisture, mold, and sprouting. Store shallots in a cool, dry, well-ventilated area and they should keep for a month or so, provided they were in good condition when purchased.

The flavor of shallot is at its best in a classic sauce such as *béarnaise*, *Bercy*, or the butter sauces *beurre blanc* and *beurre rouge*. These make fine accompaniments to simply prepared (poached, broiled) seafood, chicken, and beefsteak. Shallots are delicious, and traditional, in steamed mussels. They also flavor tomato sauces and butter for snails, and are delicious when roasted whole to buttery softness, then squeezed out of the skin and spread on toast or puréed to add body and delicate flavor to sauces for roasts and sautés. Raw shallots are very pleasant in salads. Should they begin to sprout, simply snip the tiny greens for sprinkling over salad–and feel free to use the bulb too, so long as it is free of rot and fairly firm.

WHAT TO LOOK FOR:
dry, papery greyish to reddish brown skin; firm, heavy cloves

WHAT TO AVOID:
softness, moisture, mold, and sprouting

HOW TO STORE:
cool, dry, ventilated place for a month or so

PRIME SEASON: (Indicated by darker shade).

| JAN | FEB | MAR | APR | MAY | JUN | JUL | AUG | SEP | OCT | NOV | DEC |

SORREL

Sorrel has been in the kitchen for centuries in several wild and cultivated forms. The *oseille* of France is perhaps the most visible form, but the Eastern European kitchen would not be the same without sourgrass, some varieties of which are called Dock, Sour Dock, Dock Sorrel, Curled Dock, Bitter Dock, Garden Sorrel, Round Sorrel, Virgin Sorrel (it produces no seeds), and Sheep Sorrel, to mention only the major ones. Whatever name or variety, the herb is characterized by an appealing and assertive sourness, the result of oxalic and other acids, that is much prized wherever it is widely used. Beyond Central Europe, Scandinavia has found many uses for this perennial, but sorrel has never gained a wide following in Britain or the U.S.

If you can find it, sorrel will be best when picked young in spring or early summer or, more likely, any time of the year from greenhouse crops. Most of the supply goes to restaurants and processing plants, but fancy markets do sometimes carry it. Select small leaves if possible, no more than five inches in length. Larger leaves are more pungent but still edible, though woody stems are undesirable. Look for bright green, crisp leaves. As usual with greens, avoid wilt and yellowing. If necessary, sorrel can be wrapped and refrigerated for a few days, but it is best when very fresh.

Sorrel is famous for two traditional soups: the cream soup *Potage Germiny* of France, and the Eastern European *schav*. When very young, sorrel makes a sprightly salad green, in small quantities, but it is more often cooked–steamed in its own moisture after careful washing–and served buttered, creamed, or gratinéed with cheese. Another traditional French preparation calls for stuffing shad with a sorrel purée, which allows the acids to break down the tiny bones, making that anatomically difficult fish much easier to eat. Because sorrel complements fish so well, one of the glories of the nouvelle cuisine is scallops of salmon quickly undercooked in a no-stick skillet and served with a sauce of sorrel, white wine, and cream.

WHAT TO LOOK FOR:
small to medium, bright green leaves

WHAT TO AVOID:
wilt, rot, yellowing, woody stems

HOW TO STORE:
refrigerate wrapped, unwashed, for a few days

PRIME SEASON: (Indicated by darker shade).

| JAN | FEB | MAR | APR | MAY | JUN | JUL | AUG | SEP | OCT | NOV | DEC |

SPINACH

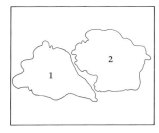

1. Semi-savoy
2. Savoy

'Gotta eat your spinach, Baby,' was the advice Shirley Temple parlayed into one of her biggest musical hits. Getting children to eat spinach was an often fruitless chore that preoccupied American parents for generations. Parents dutifully force-fed their youngsters in the hope of providing them with a diet rich in blood-building iron until the 1960s, when spinach came under attack for its oxalic acid, which can rob the body of its ability to absorb calcium if consumed in sufficient quantity. Now that the smoke has cleared, spinach is recognized to be a safe, reasonably nutritious green that can also be delicious.

Most of the spinach for fresh market is of the sort called savoy. Crinkled leaves are characteristic, and there are also partially crinkled ones, called semi-savoy, and flat-leafed varieties. They are interchangeable. Shoppers need be concerned only with the freshness of the leaves, not their shape. Spinach is in good supply all year with slightly less quantity and quality during the hot months. Choose smaller leaves that are deep-green and crisp. Avoid wilt, rot, and bruising. Thick central seed stalks indicate that the spinach is overage and unacceptable. Much of the crop is trimmed, washed, and packaged in ten-ounce plastic bags. These are all right, but bunches (called rosettes) with their pale purplish stems intact will stay fresh longer. Spinach is tastiest and most economical when very fresh, but it can be bagged and refrigerated for a day or two if necessary.

Spinach grows in sandy soil and must always be carefully washed, even if you have purchased a package that states that the contents are washed. Place the spinach in a large basin of water, agitate gently, and keep changing the water until all traces of grit are gone. If the spinach is to be cooked, one of the changes could be hot water, which will tend to relax the crinkles and release the sand. Only the really coarse stems need to be removed for most purposes, but if you feel the need you can strip the rib as well by holding the leaf shiny side up and pulling the stem down through your fingers. It will break where the rib fibers become tender. Spinach makes a refreshing salad green, especially with highly seasoned ingredients and perhaps a touch of sweetness. As a cooked vegetable, spinach is usually steamed in its own moisture just until wilted, then served with the chosen seasonings or squeezed to remove the considerable excess moisture and chopped or puréed to be combined with other ingredients for more elaborate dishes. Spinach is a major flavor in dishes called Florentine, in honor of Catherine de Medici's fondness for it, and Viroflay, for the Paris suburb that used to be famous for fine quality spinach. Wild, Chinese, and New Zealand spinaches are unrelated plants that can be used in the same ways as true spinach.

WHAT TO LOOK FOR: crisp, dark green, smaller leaves

WHAT TO AVOID: wilt, rot, bruising

HOW TO STORE: bag and refrigerate for a day or two if necessary

PRIME SEASON: (Indicated by darker shade).

| JAN | FEB | MAR | APR | MAY | JUN | JUL | AUG | SEP | OCT | NOV | DEC |

SPROUTS

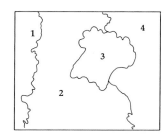

1. Mung bean
2. Alfalfa
3. Radish (Tsu-mamina)
4. Soybean

Vegetables that any fool can grow–that's what sprouts are. But wise people are the ones who are growing them. No soil, no fertilizer, no pesticides, no climate worries, no seasons! All that sprouts require is a supply of fresh water, a dark place at room temperature, and several days–low overhead to be sure. Professional growers get high yield, and amateurs get remarkably nutritious and tasty dividends from their investment.

Any edible seeds can be sprouted, though mung beans, soybeans, and alfalfa are the most popular. Radish sprouts once had a vogue among do-it-yourselfers, and these peppery little morsels are back, thanks to a seed of Japanese origin that is now being sprouted and sold under the trade name 2-mamina (a corruption of the Japanese name, best transliterated as *tsu-mamina*). Health food enthusi-

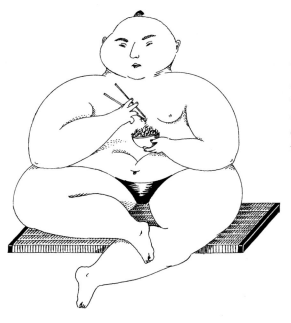

asts sprout various organically grown seeds, including wheat, barley, lentils, and corn. To sprout seeds at home, simply change the water twice a day and the growth should be ready for the table in three to five days. Alfalfa, radish, and other very tender sprouts are given some sunlight after germination to green them before use. Seeds that have been chemically treated for planting must be avoided, as must any seed that is poisonous when raw, especially lima beans and broad beans (fava).

In the market, look for sprouts that are crisp and fresh. Reject sprouts with withered ends, wilt, or rot. Radish sprouts should be very green, alfalfa green and white, and bean sprouts pale ivory or beige. All parts are edible. For nutrition's sake, do not remove roots or empty seed skins. Freshness is essential, but you may wash the sprouts, wrap them, and refrigerate them for a day or two if necessary.

Sprouts of all sorts are delicious in salads. The fat, pale bean sprouts are also used in cooking. They are most successful when they are stir-fried or used in quickly cooked egg dishes or dumplings. Of the bean sprouts, the smaller mung have a more delicate flavor than soy. Sprouted grains–wheat, corn, barley–have a pleasant sweetness that makes them prized for baking.

WHAT TO LOOK FOR:
fresh, crisp sprouts of proper color

WHAT TO AVOID:
withered ends, wilt, rot

HOW TO STORE:
wash, bag, and refrigerate for a day or two if necessary

PRIME SEASON: (Indicated by darker shade).

JAN	FEB	MAR	APR	MAY	JUN	JUL	AUG	SEP	OCT	NOV	DEC

SUMMER SQUASH

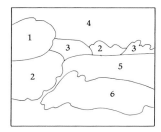

1. Chayote
2. Pattypan
3. Yellow
4. Chinese okra (*see gwa*)
5. Hairy squash (*dzeet gwa*)
6. Zucchini

Notice that the zucchini skin is a bit blemished, almost an inevitability because of its tenderness. Yellow squash is equally fragile, and you will not always find them as perfect as these.

Summer is an excellent time to find them, but spring and fall are also promising, and winter is not bad either. Summer squashes are so named for their perishability rather than any strict seasonality these days. Their most important characteristic is that they are immature fruits, young and tender and usually edible raw. All of them will mature into large, hard-skinned winter squashes, and the one that is most often allowed to do so is zucchini, or courgette, which will mature into a vegetable marrow and gain a blandness and texture that only a Briton could love.

Look for very young squashes that are fresh, crisp, and unblemished. Rot sets in quickly, so avoid bruising. Zucchini and yellow squashes, straight or crook necked, are best when under seven inches long and are finest when they are tiny. You might find Golden zucchini, or two squashes that are similar to zucchini, called Cocozelle and Caserta. Chayote and Pattypan should be very firm and no more than four inches long. The Chinese summer squashes are always larger, but hairy squash, *dzeet gwa*, should always be very hairy to ensure its youth. Summer squashes can be wrapped and refrigerated for a few days if necessary.

Zucchini and yellow squash are delicious when lightly steamed or sautéed and served with just butter or olive oil and chopped fresh herbs. They also team well with other vegetables, especially tomatoes, onions, garlic, peppers, and eggplant. They

may be fried, stuffed, stewed, gratinéed, or used cooked or raw in salads. Their blossoms, sometimes with tiny fruits attached, are stuffed or fried as fritters. Pattypans are also sometimes harvested tiny with blossoms attached for Italian stuffed or frittered preparations. More mature Pattypans, sometimes called cymling or scalloped, can be sliced and fried or used for any other summer squash purposes. Chayote, also called mirliton, vegetable pear, and a host of Caribbean names, is treated more like a winter squash, usually stuffed with a savory meat, sausage, or seafood stuffing and baked. The Chinese squashes are used in soups and stir-fries. *See gwa*, called silk squash, pleated squash, or Chinese okra, must have the tough ridges of each spine pared away before it is sliced and cooked.

WHAT TO LOOK FOR:
very young, fresh, crisp, unblemished squashes, very firm or more tender depending upon variety

WHAT TO AVOID:
wilt, cuts, bruises, browning, shriveled skin

HOW TO STORE:
wrap and refrigerate for a day or two if necessary

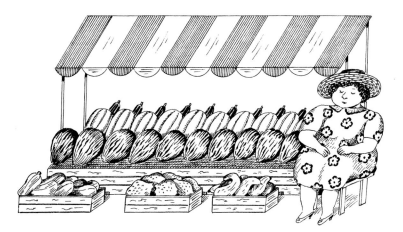

PRIME SEASON: (Indicated by darker shade).

| JAN | FEB | MAR | APR | MAY | JUN | JUL | AUG | SEP | OCT | NOV | DEC |

Winter squash

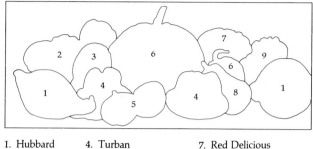

1. Hubbard
2. Acorn
3. Spaghetti
4. Turban
5. Golden Nugget or Golden Acorn
6. Pumpkin
7. Red Delicious
8. Butternut
9. Buttercup

They are planted in summer, harvested in summer or fall, and called winter. Terminology like that was never meant to confuse–it simply dates back to a time when the seasons were more crucial to man's survival than they are now. 'Good keepers' became known as winter vegetables if they could be persuaded to hold up even through December, and hard-shelled squashes do their duty well. Winter melons are another example of this kind of classification, not to be confused with Chinese winter melon, which is named for its frosty bloom (it is a winter squash). Winter squashes are mature fruits that are always cooked. Most of them have orange flesh that is slightly sweet with a nutty flavor. The seeds are often toasted to make nutritious snacks. Fall is peak season, though a few are available year round.

Look for sturdy, heavy winter squashes with fairly glossy skin that is unblemished by soft spots, cuts, breaks, or uncharacteristic discolorations. They should keep well for a week or more at cool room temperature, or for a month or two if they are stored in ideal root cellar conditions–a cool, dry, dark, well-ventilated place.

Acorn squash

This is probably the most available of them all. Acorns can be found most any day but are at their best in fall and early winter. The pale-fleshed spring and summer ones are not very good. The skin is usually deep green, though it may be streaked with gold or be very golden, in which case you may call it a Golden Acorn, although that name is also used for a related type called Golden Nugget. Acorns are usually baked, unpeeled, sometimes after parboiling, and served in halves or rings either sweetened and spiced or stuffed with a savory sausage stuffing.

Hubbards

These are huge things that make delightful ornamental curiosities, as well as good eats. The ones shown here are called blue hubbards, and there are green or orange rinds as well. After peeling, the orange flesh is baked, steamed, or simmered and served in a variety of ways. It can be sweetened and spiced or simply buttered. Purées, soups, stews, puddings, and pies are possibilities.

Spaghetti squash

This is different from the others in that when cooked tender, the pale flesh can be coaxed into long strands that resemble and can be served like spaghetti. The flavor is bland and not as satisfying as pasta, but worth a try.

Turbans

These come in bizarre shapes with extravagant coloration that makes them popular as harvest ornaments, though they can also be prepared in any way that hubbards are.

Pumpkins

These are American favorites that get their biggest use as jack-o'-lanterns. Pumpkin pie is a great favorite for holiday desserts, and any use for hubbards is appropriate, especially a soup for which a pumpkin is used as a serving tureen.

Golden delicious

This is another huge one of the hubbard sort with all the same properties.

PRIME SEASON: (Indicated by darker shade).

| JAN | FEB | MAR | APR | MAY | JUN | JUL | AUG | SEP | OCT | NOV | DEC |

BUTTERNUT SQUASH
It is one of the most popular of the winter squashes. Its size is manageable and its flesh is good and sweet for any of the hubbard uses. Butternuts are available often, though at their best in fall and early winter.

BUTTERCUPS
These are of the turban sort, but tamer in appearance with a green background color. They cook well.

CALABAZA
This is a Spanish generic term for squash that also refers to a few specific ones, especially Caribbean winter squashes that are large and more pumpkin-like than anything else. Calabazas are most available in markets that cater to Latin tastes, and they are cooked in beans, stews, and sweets.

Any summer squash may be overgrown to maturity and treated like a winter squash. The results are likely to be pale and bland, of little interest on their own – mere foils for rich or highly seasoned sauces.

WHAT TO LOOK FOR:
sturdy, heavy fruits, glossy skin

WHAT TO AVOID:
dull, shriveled skin, soft spots, cuts or breaks in the skin, uncharacteristic discolorations, mold, rot

HOW TO STORE:
a week or more at cool room temperature or for a month or more if stored in a cool, dry, dark, well-ventilated place

STAR FRUIT

You must admit it is pretty. Strange, but pretty. Star fruit, properly called carambola, is but one of the minor miracles performed by those canny produce people in New Zealand. There are probably more accurate appraisals of it all, but it seems that we owe the presence in the West of all these charming new fruits to European economic politics. In a nutshell (or a kiwi egg), Britain supported a thriving dairy industry in its former colony for many years until membership in the Common Market forced Mother England to buy from its neighbors. So New Zealand turned to lamb and fruit, with a vengeance. Star fruit is one of the happy results of this efficient combination of natural resources, transport technology, and advertising that puts Madison Avenue to shame.

Star fruit is usually available in the fall and through the winter. For shipping purposes (by air) it is picked young and firm, so it never has the tropical sweetness of tree-ripened fruit. Depth of golden color is the clue to ripeness, but even very pale star fruit has an appealing tartness that makes it refreshing. Choose plump, unblemished fruit. The five spines will probably be a bit dark and dry-looking. You may age the fruit at room temperature in a paper bag with a few holes in it, and it may mellow and ripen some, but do not expect miracles. Refrigeration is unnecessary unless you must keep star fruits for a few days beyond the point where they have become soft and fragile.

Star fruits should be carefully trimmed to remove just the tough, darkened edges of the five segments. Usually star fruits are sliced crosswise to reveal the five-pointed star pattern. The ends and the seeds are discarded. Most often in the West these costly little gems are used to add an exotic touch to cold food presentations. They can also be used for purposes where other tropical fruits are welcome – jams, jellies, preserves, pickles, relishes, and even beverages and hot dishes.

WHAT TO LOOK FOR:
plump, unblemished fruit (expect dark spines) of pale or deep yellow

WHAT TO AVOID:
mushiness, bruising, rot

HOW TO STORE:
mature in a paper bag at room temperature, refrigerate for a day or two if very soft

PRIME SEASON: (Indicated by darker shade).

| JAN | FEB | MAR | APR | MAY | JUN | JUL | AUG | SEP | OCT | NOV | DEC |

Sweet potatoes

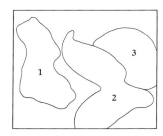

1. *Batata blanca*
2. Red sweet potato
(North Carolina yam)
3. White sweet

Potato is their rightful name. Before the white thing took over world interest in the nineteenth century, potato, from the Caribbean *batata*, was applied to sweet potatoes without any modifier. The round potato from Peru adapted itself to soil conditions and eating habits around the world and left sweets out in the cold, or, more accurately, in the heat of the sunny climates they require. Luckily, they were never squeezed out entirely and remain plentiful in the Western Hemisphere and the Pacific islands. Sweet potatoes never really caught on in Europe, though there are some white ones grown in Spain, Portugal, and Italy. In Africa and most of Asia native yams are more important, and this brings us to the big sweet potato/yam confusion, which is centuries old. Yams and sweet potatoes are tubers that happen to be similar in appearance and flavor, but they are from totally different plants. The name yam applied to American sweet potatoes dates back to African slaves who assumed they had found a familiar vegetable in the New World.

Sweet potatoes are grown in several varieties, especially the richly colored orange ones with red-brown skin (these are the ones called yams), the white-fleshed ones that look like a paler version of the orange, and the white ones called in the Caribbean *batatas blancas*, which have a distinctive red skin and fat, well-tapered shape. They are available all year, but are at their best and most plentiful in fall and early winter. Look for firm, unblemished sweet potatoes that are smooth, bright, and not too misshapen. Small to medium-sized tubers are usually tenderer and sweeter than overgrown ones. Avoid dull, dry, shriveled sweet potatoes and any bruising, for rot spreads quickly. Sweet potatoes should be stored in ideal conditions for roots–a cool, dry, dark, well-ventilated place. Without such a setup, buy only what you will use in a week's time and store them at cool room temperature. Allow for air circulation.

The white sweet potatoes can be used in recipes for white potatoes, though they can also be used like the orange ones. Sweet potatoes are baked, fried, candied, mashed, puréed, or made into biscuits, breads, fritters, or soufflés. Sweet potato pie is a traditional Southern favorite. Orange juice or zest goes well with sweet potatoes, as do rum, bourbon, ginger, sweet spice, dark sugars and syrups, pecans, and raisins or currants.

WHAT TO LOOK FOR:
small to medium size; firm, unblemished, smooth, bright; not too misshapen

WHAT TO AVOID:
dull, dry, shriveled skin; bruising

HOW TO STORE:
cool, dry, dark, well-ventilated place or cool room temperature for about a week

PRIME SEASON: (Indicated by darker shade).

| JAN | FEB | MAR | APR | MAY | JUN | JUL | AUG | SEP | OCT | NOV | DEC |

TOMATOES

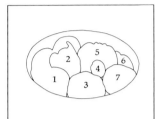

1. Standard commercial
2. Green
3. Local crop (New Jersey)
4. Cherry
5. Holland beefsteak
6. Plum (egg)
7. Hothouse

Note the variety of shapes, sizes, and colors. Only the locally grown tomato and the green tomato (for cooking) are likely to be of fine quality. More and more imports from Israel, Holland, and Spain are available during cold weather; while they are beautiful, juicy, and far superior to standard supermarket fare, nothing can compete with local summer tomatoes.

It is difficult to imagine an edible generating such controversy, but tomatoes have had the most checkered career of all. Early Spanish explorers found them in Mexico and brought them back to Spain to a cautious reception. Then botanists recognized them as cousins to belladonna and deadly nightshade (along with potatoes, peppers, eggplant, and tobacco) and pronounced them poisonous, or just loathsome. In typical fashion, the French decided the new arrival must be an aphrodisiac and named it *pomme d'amour,* 'love apple.' To this day, in that dying civility called formal dining, it is considered poor taste, vaguely obscene, to serve a tomato in recognizable form. Only in Naples did it find a happy home, where the little yellow thing named *pomo d'oro,* 'golden apple,' was enthusiastically bred, leading to the tomato in the forms we know today.

Only in the twentieth century have tomatoes gained worldwide acceptance, and if you think they had a rough time in the beginning, consider what is happening to them now. The modern commercial tomato is hard, thick skinned, and tasteless. Despite the industry term 'vine ripened,' these tomatoes are actually picked green and put into cold storage, a practice that destroys their ability to ripen. They will soften and turn red, especially if gassed with ethylene, but they will not develop characteristic sweetness, aroma, and vitamins. Only locally grown summer fruits are superior, though there is some flavor to be found in cold weather from cherry tomatoes and the beautiful imports from Israel. Choose fruit that is plump and fragrant, yielding to

gentle pressure. Red tomatoes should be very red with no green showing; yellow ones should be very yellow and, of course, green tomatoes for cooking should be very green and firm. Honest local tomatoes that have been picked at or near maturity are likely to be a bit ugly and irregular compared with the supermarket product, so welcome the difference. Finish ripening at room temperature in a paper bag with a few holes in it.

Tomatoes have myriad kitchen uses, raw and cooked. They are used for relishes, ketchups, soups, even cakes, and sauces of all sorts, hearty and delicate, for meats, poultry, fish, pasta, and other vegetables. They may be baked, broiled, sautéed, fried, stuffed, stewed, or made into sherbet or ices. Tomatoes are used in many salads and sandwiches. In fact, they go with almost anything. Tradition dictates different types for different purposes, but the plum-shaped ones usually reserved for cooking are excellent raw. For winter cooking, the best choice is canned Italian plum tomatoes.

WHAT TO LOOK FOR:
plump and fragrant, yielding to gentle pressure, proper color

WHAT TO AVOID:
bruises, cuts, extensive scarring, insect blight, flabbiness, poor color, serious growth cracks, rock-hard fruit (except green)

HOW TO STORE:
ripen at room temperature, then refrigerate for a day or two if necessary; never refrigerate underripe fruit

PRIME SEASON: (Indicated by darker shade). This may vary slightly due to local crops.

| JAN | FEB | MAR | APR | MAY | JUN | JUL | AUG | SEP | OCT | NOV | DEC |

TRUFFLES

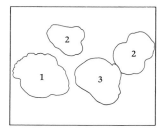

1. Italian black
2. Italian white
3. French black

The white truffle on the right has begun to show a little moisture, a result of age or chill damage. It is still usable so long as it is firm and not actually rotting, but it is unacceptable at usual truffle prices.

Subterranean fungus. The notion does not exactly inspire confidence, yet for centuries people have been willing to spend a king's ransom for just such a thing, so long as it is a truffle from France or Italy. Truffles are legendary quirks of nature that cannot be reliably controlled or cultivated. They simply grow where the conditions are right, usually in the soil around the roots of oak trees, and it is up to man, with the help of beast, to sniff them out. Dogs and pigs are quite good at it. Truffle hunters have their own preferences, mostly dogs these days, because when checking out secret territories it is easier to feign a casual stroll with a pet dog than a pet pig. And as one veteran put it, 'Pigs are, well, pigs.' Mammals love truffles. In fact, they are sexy—truffles release an aroma that is chemically identical to certain scents released by male animals. It is a heady scent, both earthy and sublime.

There are truffles growing in many parts of the world, but the important ones are the black truffles from France and black and white truffles from Italy. The season is usually late fall for white and winter for black. These are the ones that fetch the highest prices, though you might find others—grey, brownish, or reddish, even summer truffles—that are less interesting and less costly. A truffle in good condition will be plump, firm, and dusty-looking. Avoid truffles that are soft or dessicated and rock-hard. When over-the-hill they may also look gooey. Though perishable, truffles should keep for about a week if they are buried in rice in a closed container and refrigerated. Truffles should be brushed with a soft brush, not washed, to remove surface dirt, and cut truffles can be stored in a sealed jar, covered with Madeira or white wine, and refrigerated.

Black truffles are always cooked to bring out their flavor. They are used in a wide variety of recipes, especially *sauce périgueux*, *poularde demi-deuil* ('chicken in half-mourning,' so named for the black that truffle adds), and other poultry, game, or beef preparations, truffles in puff pastry, pâtés, terrines, foie gras, eggs and omelets, and even sautéed potato cakes. White truffles are not cooked, but rather grated into fine shavings onto the top of the finished dish, most likely pasta or risotto. Never discard the rice used for packing and storage, because it has been infused with the perfume of its charge and is delicious used in cooking. Canned truffles have poor flavor and are still expensive to use.

WHAT TO LOOK FOR:
plump, firm, dusty-looking

WHAT TO AVOID:
soft, dessicated, or rock-hard truffles, moist spots

HOW TO STORE:
bury in rice in a closed container and refrigerate for about a week

PRIME SEASON: (Indicated by darker shade).

| JAN | FEB | MAR | APR | MAY | JUN | JUL | AUG | SEP | OCT | NOV | DEC |

TURNIPS AND RUTABAGAS

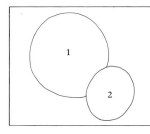

1. Rutabaga
2. Turnip

Note the little sprout on the turnip, a common flaw that betrays age. This one is barely acceptable, and should be rejected in favor of fresher turnips if possible.

A little arsenic is good for the body; otherwise turnips would not have been rustic staples since prehistory. Their peasant reputation is probably the major reason that turnips rank high on the list of vegetables that people say they don't like very much, and tough, woody, overgrown, strong-flavored winter turnips can deter all but the most ardent fans. Spring is the season for fresh, young turnips, when they are mild and tender and surprisingly sweet. Otherwise, the greens are still good, even if the roots leave something to be desired. A close relative of white turnips is the big, hard, yellow-fleshed thing called rutabaga or swede. This one is so unpopular that it is nearly always waxed to extend its shelf life for months. The rutabaga in the market today may be the one there next month unless you buy it and enjoy it now. Reasonably fresh rutabagas have a delightfully earthy flavor that children hate and adults should try again.

Fall and winter are peak seasons for turnips and rutabagas, though turnips are best in spring.

Choose small roots that are plump and smooth. If greens are attached, they should be fresh. Avoid withered, lightweight, or overgrown specimens and bruising. If the tops have been trimmed there should be no sprouting. Rutabagas hold best under root cellar conditions–cool, dark, dry, well-ventilated–but they can be refrigerated, as can turnips, for a week or so.

Turnips and rutabagas are delicious when they are roasted or sliced and baked with butter (parboil the rutabagas). Turnips are especially good for flavoring in soups like the classic French *pot-au-feu*, or in stews like the traditional springtime ragout of mutton called *navarin printanier*. Spring turnips are also served with duck. Turnips and rutabagas both may be puréed, mashed, creamed, baked into a pudding, combined with other vegetables, or used raw in salads or slaws. Rutabaga greens are famine food, but turnip greens, which are essentially the same thing as rape, are fine and are especially popular in Italy and the American South.

WHAT TO LOOK FOR: small, plump, smooth, fresh greens, if any

WHAT TO AVOID: withered, blemished, lightweight, or overgrown roots; sprouting; stale greens

HOW TO STORE: refrigerate for up to a week, or store rutabagas in a cool, dark, dry, well-ventilated place for a month or more. Trim greens, if any, before storage

PRIME SEASON: (Indicated by darker shade).

| JAN | FEB | MAR | APR | MAY | JUN | JUL | AUG | SEP | OCT | NOV | DEC |

Watercress

If salad greens can be said to strike various notes, or chords, or tones, then watercress is one of the sprightliest music-makers of them all–a piccolo perhaps? *Peppery* is the usual adjective, and it will do nicely. Nothing can perk up a salad with more finesse than this pretty little member of the mustard family. One curious property of watercress observed and recorded through the ages is that while it tastes hot, the sensation actually produced in the mouth is cooling and thirst-quenching. The ancients would have assumed that watercress brought the essence of its cool aquatic life (yes, it really grows in water) from limestone spring to table. Though this explanation is short on accuracy, it is long on charm and quite sufficient. Watercress has another life as a cooked vegetable, especially in soups, be they French or Chinese. The application of even the slightest heat changes it drastically, eliminating the pepper and leaving a soft green herbaceousness that is most satisfying.

Fresh watercress is available year round with a smaller supply in winter. Look for bunches of crisp, deep-green leaves that are free of any wilt, rot, bruising, or yellowing. Large woody stems are not especially valuable in the kitchen, so you can try to select bunches with smaller, tenderer stems. Watercress is highly perishable, so it should be used the day of purchase if at all possible. Otherwise, rinse with cold water to moisten it, wrap closely, and refrigerate for a day or two. For even more efficient keeping, stand the stems in a glass of water and wrap the leaves, sealing all around the glass. Refrigerated this way, watercress stands a better chance of staying crisp and green.

Watercress is most often used in salads, where its peppery flavor makes a fine complement to nearly any other ingredient. It is also an attractive and valuable platter garnish for all manner of foods, especially successful with cold meats and poultry. The Chinese stir it into hot clear soups just at the last moment, and they use it in stir-fried dishes as well. The French create rich but delicate cream soups based on watercress, alone or in combination with other greens. There are also watercress sauces to be made. Though rarely used this way, it makes a fine potherb. In most cases, only the leaves and tender little stems are used, though some people like to munch on the crunchy large stems.

WHAT TO LOOK FOR:
crisp, deep-green leaves

WHAT TO AVOID:
wilt, rot, bruising, yellowing

HOW TO STORE:
refrigerate with the stems in a glass of water or chopped ice and the leaves tightly wrapped, a day or two

PRIME SEASON: (Indicated by darker shade).

| JAN | FEB | MAR | APR | MAY | JUN | JUL | AUG | SEP | OCT | NOV | DEC |

Watermelon

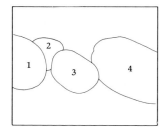

1. Crimson Sweet
2. Sugar Baby
3. Yellow
4. Jubilee

In simpler times, fruits and vegetables came and went with the seasons. It was believed, with good reason, that this orderly progression of things had value for man's soul as well as for his belly. No clearer example of healthy seasonality exists than watermelon, the ideal summer treat. With impeccable timing, watermelons reach their perfection just when the long, sweltering days make the cool, inviting, juicy sweetness a godsend. So thirst quenching are they that watermelons are cultivated in desert and semitropical areas of their native Africa as a water source in times of drought. In modern fashion, watermelons can now be had any time of the year, given the right purse. But who in the world would want to eat watermelon in winter?

The best watermelons and the largest supply are available in June, July, and August. Tasting is the best test, but there are a few guidelines that can help in choosing a good watermelon. First, it should be large, firm, symmetrical, and very heavy. Skin color should be right for the variety and fresh-looking, with a waxy bloom that gives way to gloss. Check the underside that grew against the ground–it should be creamy or yellowish, not white or pale green. Thumping will not tell you much. Eyeing or tasting cut melon from the seller's supply should give you a fair idea of the quality of the others. The cut flesh should be deeply colored and crisp with mature seeds. Most watermelons have dark seeds, so there should be few or no immature white ones. Avoid cracks, soft or watery bruises, and flesh that is mealy or water-soaked.

Watermelons are perishable and do not improve once they are harvested. Refrigerated whole, a watermelon should last for several days, but cut pieces should be used promptly, within a day or two at most.

Eaten just as they are in wedges, chunks, or balls, watermelons provide refreshment that is low in calories and surprisingly nutritious. There are considerable amounts of vitamins A and C and even iron. Because of their size and shape, watermelons make dramatic vessels for holding fruit salad or summer beverages. They can be cut or carved intricately. Watermelon juice has been used for ices, but it is a bit bland to stand on its own. The seeds make a tasty snack when toasted, and the rind makes a delicious sweet pickle.

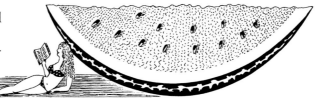

WHAT TO LOOK FOR:
large, firm, symmetrical, very heavy fruit, fresh-looking skin of proper color, underside creamy or yellowish, flesh deeply colored

WHAT TO AVOID:
cracks, soft or watery bruises, white or pale green underside, flesh mealy or water-soaked, immature seeds

HOW TO STORE:
refrigerate whole for several days or in wrapped cut pieces for a day or two

PRIME SEASON: (Indicated by darker shade).

| JAN | FEB | MAR | APR | MAY | JUN | JUL | AUG | SEP | OCT | NOV | DEC |

WHAT'S GREAT THIS MONTH

JANUARY	FEBRUARY	MARCH	APRIL	MAY
Avocados	Avocados	Artichokes	Artichokes	Artichokes
Broccoli	Broccoli	Asparagus	Asparagus	Asparagus
Brussels sprouts	Brussels sprouts	Avocados	Avocados	Avocados
Cabbage	Cabbage	Beets	Beans	Basil
Chinese cabbage	Chinese cabbage	Broccoli	Beets	Beans
Cardoons	Caulifower	Cabbage	Broccoli	Beets
Caulifower	Celery root	Chinese cabbage	Cabbage	Berries
Celery root	Cherimoyas	Caulifower	Chinese cabbage	Broccoli
Cherimoyas	Chicory	Celery root	Caulifower	Cabbage
Chicory	Fennel	Chicory	Celery root	Chinese cabbage
Fennel	Grapefruit	Chives	Chicory	Chives
Grapefruit	Ugli fruit	Dandelion	Chives	Cucumbers
Ugli fruit	Greens	Fennel	Cucumbers	Dandelion
Greens	Leeks	Grapefruit	Dandelion	Fiddleheads
Leeks	Mâche	Ugli fruit	Horseradish	Horseradish
Mâche	Mandarin oranges	Greens	Leeks	Head lettuce
Wild mushrooms	Sweet oranges	Horseradish	Head lettuce	Leaf lettuce
Mandarin oranges	Parsnips	Leeks	Leaf lettuce	Mangoes
Sweet oranges	Pears	Head lettuce	Okra	Okra
Parsnips	Rhubarb	Leaf lettuce	Sweet oranges	Sweet oranges
Pears	Shallots	Mâche	Papayas	Papayas
Prickly pears	Spinach	Okra	Peas	Peas
Shallots	Truffles	Seville oranges	Chile peppers	Chile peppers
Spinach		Sweet oranges	Sweet peppers	Sweet peppers
Sweet potatoes		Parsnips	Rhubarb	Rhubarb
Truffles		Peas	Shallots	Shallots
		Rhubarb	Spinach	Sorrel
		Shallots	Summer squash	Spinach
		Spinach	Turnips	Summer squash
				Turnips

JUNE	JULY	AUGUST	SEPTEMBER	OCTOBER
Apricots	Apricots	Apricots	Arugula	Apples
Arugula	Arugula	Arugula	Beans	Beets
Basil	Basil	Basil	Beets	Broccoli
Beans	Beans	Beans	Berries	Brussels Sprouts
Beets	Beets	Beets	Cabbage	Cabbage
Berries	Berries	Berries	Chinese cabbage	Chinese Cabbage
Cabbage	Cherries	Cherries	Cauliflower	Cauliflower
Cherries	Corn	Corn	Corn	Celery root
Chives	Cucumbers	Cucumbers	Cucumbers	Chicory
Cucumbers	Mangoes	Dates	Dates	Cranberries
Dandelion	Melons	Figs	Figs	Cucumbers
Leaf lettuce	Okra	Grapes	Grapes	Dates
Mangoes	Peaches	Mangoes	Head lettuce	Fennel
Melons	Chile peppers	Melons	Leaf lettuce	Figs
Okra	Sweet peppers	Okra	Mangoes	Grapes
Papayas	Plums	Peaches	Melons	Kumquats
Peas	Sorrel	Chile peppers	Wild mushrooms	Leeks
Chile peppers	Summer squash	Sweet peppers	Okra	Head lettuce
Sweet peppers	Tomatoes	Plums	Pears	Leaf lettuce
Plums	Watermelon	Sorrel	Chile peppers	Wild mushrooms
Sorrel		Summer squash	Sweet peppers	Nuts
Spinach		Tomatoes	Plums	Okra
Summer squash		Watermelon	Prickly pears	Pears
Watermelon			Shallots	Chile peppers
			Sorrel	Sweet peppers
			Summer squash	Persimmons
			Tomatoes	Pomegranates
			Watermelon	Prickly pears
				Quince
				Shallots
				Spinach
				Winter squash
				Star fruit
				Sweet potatoes

November

Apples
Broccoli
Brussels sprouts
Cabbage
Chinese cabbage
Caulifower
Celery root
Chicory
Cranberries
Cucumbers
Dates
Fennel
Grapes
Greens
Kumquats
Leeks
Head lettuce
Leaf lettuce
Mâche
Wild mushrooms
Nuts
Okra
Mandarin oranges
Pears
Chile peppers
Sweet peppers
Persimmons
Pomegranates
Prickly pears
Quince
Shallots
Spinach
Winter squash
Star fruit
Sweet potatoes
Truffles

December

Apples
Broccoli
Brussels sprouts
Cabbage
Chinese cabbage
Cardoons
Cauliflower
Celery root
Cherimoyas
Chicory
Cranberries
Dates
Fennel
Grapefruit
Greens
Kumquats
Leeks
Mâche
Wild mushrooms
Nuts
Mandarin oranges
Sweet oranges
Parsnips
Pears
Persimmons
Pomegranates
Prickly pears
Quince
Shallots
Spinach
Winter squash
Star fruit
Sweet potatoes
Truffles

What's Great All Year

Bananas
Carrots
Celery
Coconuts
Eggplant
Garlic
Ginger
Herbs
Kiwi fruit
Lemons
Limes
Chinese melons
Cultivated mushrooms
Onions
Parsley
Snow peas
Pineapples
Potatoes
Radishes
Roots
Scallions
Sprouts
Watercress

INDEX